Stealing the Sword

Limiting Terrorist Use of Advanced Conventional Weapons

James Bonomo

Giacomo Bergamo

David R. Frelinger

John Gordon IV

Brian A. Jackson

Prepared for the Department of Homeland Security

RAND Homeland Security

A RAND INFRASTRUCTURE, SAFETY, AND ENVIRONMENT PROGRAM

The research described in this report was sponsored by the U.S. Department of Homeland Security, Science and Technology Directorate, Office of Comparative Studies, under the auspices of the Homeland Security Program within RAND Infrastructure, Safety, and Environment (ISE), a division of the RAND Corporation.

Library of Congress Cataloging-in-Publication Data

Stealing the sword : limiting terrorist use of advanced conventional weapons /
   James Bonomo ... [et al.].
      p. cm.
   "MG-510"—Back cover.
   Includes bibliographical references.
   ISBN-13: 978-0-8330-3965-1 (pbk. : alk. paper)
   1. Weapons systems. 2. Arms control. 3. Terrorism—Prevention. 4. Terrorism—
United States—Prevention. I. Bonomo, James. II. Rand Corporation.

   UF500.S84 2007
   363.325'16—dc22

                                                                    2006017948

The RAND Corporation is a nonprofit research organization providing objective analysis and effective solutions that address the challenges facing the public and private sectors around the world. RAND's publications do not necessarily reflect the opinions of its research clients and sponsors.

RAND® is a registered trademark.

Cover design by Peter Soriano

Published 2007 by the RAND Corporation
1776 Main Street, P.O. Box 2138, Santa Monica, CA 90407-2138
1200 South Hayes Street, Arlington, VA 22202-5050
4570 Fifth Avenue, Suite 600, Pittsburgh, PA 15213-2665
RAND URL: http://www.rand.org/
To order RAND documents or to obtain additional information, contact
Distribution Services: Telephone: (310) 451-7002;
Fax: (310) 451-6915; Email: order@rand.org

Preface

In this document, we focus on how the United States can shape the environment, including the perceptions of terrorists, to discourage the use of advanced conventional weapons. We review weapons under development, assess prospective and previous terrorist uses of such weapons, identify ways to make particular kinds of weapons less attractive to terrorist groups, and explore reasons that terrorist groups choose or reject certain weapons.

The analyses presented here should be of interest to homeland security policymakers who need to understand the threat posed by advanced conventional weapons. Those concerned with developing security or defensive systems can allocate research and development and technology funding to countermeasures and defense systems with the greatest possible potential payoff. Those concerned with training security forces can adjust their curricula and concepts appropriately. And those interested in limiting the access of terrorists to advanced weapons can learn where to focus their efforts. Overall, these efforts should influence terrorist decisionmaking, deterring their use of particular weapons. Related RAND Corporation publications include the following:

- Brian A. Jackson, John C. Baker, Peter Chalk, Kim Cragin, John V. Parachini, and Horacio R. Trujillo, *Aptitude for Destruction*, Vol. 1: *Organizational Learning in Terrorist Groups and Its Implications for Combating Terrorism* (MG-331-NIJ, 2005)
- Brian A. Jackson, John C. Baker, Peter Chalk, Kim Cragin, John V. Parachini, and Horacio R. Trujillo, *Aptitude for Destruction*,

Vol. 2: *Case Studies of Organizational Learning in Five Terrorist Groups* (MG-332-NIJ, 2005)

- Kim Cragin and Sara A. Daly, *The Dynamic Terrorist Threat: An Assessment of Group Motivations and Capabilities in a Changing World* (MR-1782-AF, 2004)
- James S. Chow, James Chiesa, Paul Dreyer, Mel Eisman, Theodore W. Karasik, Joel Kvitky, Sherrill Lingel, David Ochmanek, and Chad Shirley, *Protecting Commercial Aviation Against the Shoulder-Fired Missile Threat* (OP-106-RC, 2005).

This monograph is one component of a series of studies examining the technology competition between security organizations and terrorist organizations, a critical battleground in the war against terrorism. This series focuses on understanding how terrorist groups make technology choices and how they respond to the technologies deployed against them. This research was sponsored by the U.S. Department of Homeland Security, Science and Technology Directorate, Office of Comparative Studies.

The RAND Homeland Security Program

This research was conducted under the auspices of the Homeland Security Program within RAND Infrastructure, Safety, and Environment (ISE). The mission of ISE is to improve the development, operation, use, and protection of society's essential physical assets and natural resources and to enhance the related social assets of safety and security of individuals in transit and in their workplaces and communities. Homeland Security Program research supports the Department of Homeland Security and other agencies charged with preventing and mitigating the effects of terrorist activity within U.S. borders. Projects address critical infrastructure protection, emergency management, terrorism risk management, border control, first responders and preparedness, domestic threat assessments, domestic intelligence, and workforce and training.

Questions or comments about this monograph should be sent to the project leader, Brian Jackson (Brian_Jackson@rand.org). Information about the Homeland Security Program is available online (http://www.rand.org/ise/security/). Inquiries about homeland security research projects should be sent to the following address:

Michael Wermuth, Director
Homeland Security Program, ISE
RAND Corporation
1200 South Hayes Street
Arlington, VA 22202-5050
703-413-1100, x5414
Michael_Wermuth@rand.org

Contents

CHAPTER THREE

What Advanced Conventional Weapons Are Potentially Most

Figures

Tables

Summary

This book examines one manifestation of the general technical competition between terrorist groups and security organizations—the balance between the potential use by terrorists of advanced conventional weapons and the responses available to deter or counter them. Our use of the term *advanced conventional weapons* is inclusive and broad: any new or unusual conventional weaponry developed for ordinary military forces. Such weaponry seems a priori likely to be particularly threatening in the hands of terrorists. All weaponry is obviously designed to do damage, but new design features might enable new, or at least unfamiliar, terrorist attacks. At the same time, the usual limitation of weaponry to militaries implies that various controls could be applied, albeit less stringently than controls imposed upon nuclear, chemical, or biological weapons. Consequently, the competition involving advanced conventional weaponry seems both complex and potentially important.

One example of this competition has received much attention—the balance between terrorist use of man-portable air defense systems (MANPADS) and U.S. responses. The November 2002 attacks in Mombasa, Kenya, using Russian-built MANPADS against an Israeli airliner, demonstrated that terrorists are able to acquire and use that type of advanced weaponry.[1] In response, the United States has negotiated a multinational agreement that calls for imposing both technical and procedural use controls on new MANPADS through an expansion

---

[1] Bayles (2003).

in scope of the Wassenaar Arrangement.[2] The United States has also started a pilot program within the Department of Homeland Security to demonstrate technical countermeasures suitable for protecting commercial aircraft from MANPADS.[3] But MANPADS are only one of a long list of advanced conventional weapons that are potentially attractive to terrorists. This monograph explores a range of other weapons, both those still under development and those already available but relatively unused by terrorists. The monograph identifies those weapons that require greater attention from U.S. homeland security decisionmakers and outlines a number of actions that can mitigate the use of these weapons by terrorists.

Key Weapons of Concern

This project identified five types of advanced conventional weapons that could, in the absence of mitigating measures, provide terrorists with a qualitatively new and different capability. Each of these weapon types threatens in some sense to change the nature of terrorist attacks:

- sniper rifles and associated instrumentation
- improved squad-level weapons of several types
- long-range antitank missiles
- large limpet mines
- precision indirect fire systems.

Sniper rifles and especially their electronic support equipment allow a relatively unskilled marksman a reasonable chance of assassinating an individual from great ranges—up to 2 km—which can be well outside the area that a security force guarding an official would consider threatening.

---

[2] Wassenaar Arrangement on Export Controls for Conventional Arms and Dual-Use Goods and Technologies (2003).

[3] U.S. Department of Homeland Security (2004).

Improved squad-level weapons could provide a terrorist assault force with a variety of new abilities, from individual indirect fire to the ability to eliminate a strong point with a short-range, antitank weapon. Advanced armor-piercing ammunition is available for many rifles and will easily penetrate standard body armor.

Long-range, antitank weapons can destroy any vehicle and kill its occupants from beyond 2 km. These same weapons can also destroy a small building or speaking platform. Advanced versions of these weapons are further reducing demands on the operator, which may make these weapons highly attractive to terrorists.

Large limpet mines attached to a ship's hull have the capability to sink large, oceangoing vessels. Even smaller, more common limpet mines can sink small ships; in fact, if multiple mines are carefully emplaced, these small mines can also sink large ships. In particular, cruise ships and ferries would be vulnerable to such devices, placing their many passengers at risk. Such external mines would, of course, not be detected during conventional cargo and passenger inspections.

Finally, precision indirect fire systems—primarily advanced mortars—can enable a wide range of new terrorist attacks: on crowds in outdoor venues; on valuable physical targets, such as refineries or aircraft; and on officials or other individuals appearing at known locations, particularly in the open, such as at a press conference.

In all of these five cases, the new systems could enable the attackers to surprise security forces. The attacks could come from far beyond any controllable security perimeter, could allow a high probability of escape for the terrorists, or could require only a single, small attack to be effective.

Reducing the Threat by Raising Awareness

The first step in limiting the threat from these systems is to raise awareness of the threat. In all cases, key groups need to understand the capabilities provided by these systems. Awareness of the new capabilities should allow technical or operational changes by security forces. Such efforts may include the following key groups and threat mitigation measures:

- Personal protective services, such as the U.S. Secret Service, whose job it is to guard high-profile individuals, need to realize that snipers and antitank weapons can make lethal line-of-sight strikes from over 2 km away. They also need to realize that non–line-of-sight weapons, such as precision mortars, will soon allow very long-range, precise attacks on targets at known locations. This awareness should enable protective services to reduce opportunities for terrorist to make use of such weapons.
- Guard forces at sites and facilities need to be aware of the capabilities that new, squad-level weapons would provide to an assault force attacking them. For example, the addition of precise, indirect fire grenades should generate a greater concern with overhead cover. New rocket-propelled grenades, thermobaric warheads, and short-range antitank weapons will require enhanced fortification at strong points. Even today, currently available small-arms ammunition should motivate upgrades in guard forces' personal armor.
- Operators of cruise ships and ferries—particularly oceangoing ferries—should be aware of the potential use and impact of large limpet mines. This awareness should motivate the use of protective cordons and hull inspections before leaving port.

Reducing the Threat Through Procedural and Technical-Use Controls

Beyond awareness are procedural and technical-use controls. Most advanced conventional weapons are intended only for military use. This means that basic procedural controls governing the use of military systems will provide some limits on terrorist uses. We note two major exceptions not under such controls: sniper rifles and their accessories and advanced ammunition. For those weapons, only awareness and the precautions taken by security forces can mitigate their threat.

All the other advanced systems will presumably be subject to international procedural controls common to military systems; these controls likely will slow their diffusion to terrorist groups. But, as the preceding example of MANPADS clearly shows, even relatively expensive, controlled systems can end up in terrorists' hands.

Adding technical-use controls in many instances would represent a major step—both organizationally and technologically. First, to be effective, such technical controls require an international agreement. The continuing, complex diplomatic efforts to enhance the controls over MANPADS, where the threat has already been clearly demonstrated, illustrate the scale of any new diplomatic effort that would be required concerning other advanced weapons. We believe that to justify and to motivate such an effort would require both an increased awareness of the threatening weapon system and also readily implementable technical controls. In our view, most of the systems do not meet these two criteria.

One system, we assert, does meet both of the criteria—precision, indirect fire systems based on an advanced mortar. Many terrorists already have had some favorable experiences with mortars, notably including those terrorists being trained in the ongoing Iraqi insurgency. Because future advanced mortar systems must depend on the Global Positioning System (GPS) or an equivalent satellite system,[4] these precision, indirect fire systems also have technical features that could facilitate various sorts of use controls. In particular, integrated electronic systems involving GPS can be designed to require a "trusted component," which would be difficult for a terrorist group to circumvent. At the same time, this trusted component would serve as the key element for technical controls. A range of limitations then becomes feasible in principle, such as the imposition of expiration times or geographic boundaries beyond which the system would not function. Importantly, these limits would be all but invisible to legitimate military users, so they would add little operational burden. But the intent

---

[4] Since GPS is the only widely available satellite navigation system today, the examples and discussion in this book all refer to GPS. The arguments we make involving GPS would hold for any similar satellite navigation system, however.

of these limitations would be to make them unreliable and unattractive to most terrorist groups, particularly as unauthorized users would have no way of checking the precise times, places, or circumstances in which the system would fail.

The best time to implement such technical controls is when the system is in its design phase. Controls added "on top," after a design is "frozen," tend to be easier to circumvent. Fortunately, the most threatening system we have identified—the GPS-guided mortar without terminal guidance—is not yet in its development phase. This situation creates an opportunity to consider ways to apply the appropriate controls. We expect that this window of opportunity will close within the next few years, however, because the military utility of and demand for such a system will be high.

Steps for Moving Forward

The most worrisome advanced conventional weapons that we have identified in this research are advanced, GPS-guided mortars. Only these systems combine a significant, new capability for terrorists with a lack of effective operational counters for security forces. We must take advantage of a fleeting opportunity to design controls into the weapons. This means that starting efforts to control advanced mortars now is urgent. Although seemingly less threatening, the other advanced weapons—sniper weapons, advanced small arms, antitank guided weapons, and limpet mines—still do require some responses. Most important, they require simple awareness on the part of security forces, and also some new techniques, such as external searches of ships before leaving port.

If the United States chooses to pursue opportunities to place additional procedural and technical-use controls on precise, indirect fire weapons, such as GPS-guided mortars, we believe that two initial steps are called for. The first step is to begin diplomatic discussions with the key producer nations, so that all the involved decisionmakers and stakeholders begin evaluating potential terrorist uses of these systems. The second step is to commission a detailed study of the technical modules and architecture needed to implement proposed technical controls. Such an investigation would be directed at determining whether the existing technical modules would be sufficient or whether they might need to be modestly expanded to include the required control functions.

The U.S. Department of Homeland Security can play a key role in both these steps. Regarding the first step identified above, the department has the primary responsibility for deterring terrorist attacks. It could use that role, within the interagency process, to push for starting diplomatic discussions. This may also entail changes in the interagency system, such as permanently including the Department of Homeland Security on interagency panels that are considering arms exports. For the second step, the department could itself directly fund such a study, perhaps in concert with the U.S. National Security Agency.

While there appears to be sufficient time to negotiate and develop meaningful controls on GPS-guided mortars, that opportunity can be lost if the United States does not begin the process soon. Missing this opportunity would reduce the controls on these mortars to the existing procedural ones for military systems in general and so increase the burden on security forces to plan around and counter such attacks. Although that may be a sufficient response for the other weapon systems we have analyzed, it appears to us to be insufficient for limiting the threat from these future, advanced mortars.

Acknowledgments

The authors appreciate the thoughtful comments from our client, Robert Ross; from the program director, Michael Wermuth; and from our reviewers, James Chow and Peter Zimmerman. The resulting document, of course, remains the sole responsibility of the authors.

Abbreviations

AGLS	Automatic Gun-Laying System
ALFO	Light Fiber-Optic Weapon (Armement Léger à Fibre Optique)
AICW	Australian Advanced Individual Combat Weapon
ATGW	antitank guided weapons
ATK	Alliant Techsystems
BAC	Bindon Aiming Concept
CBRN	chemical, biological, radiological, and nuclear
CEP	circular error probable
CTD	concept technology demonstrator
DARPA	Defense Advanced Research Projects Agency
DSTO	Defence Science and Technology Organisation
ER-DPICM	extended-range dual-purpose improved conventional munition
ERMC	extended-range mortar cartridge
FAQs	frequently asked questions
FY	fiscal year
GLS	gun-laying system

GPS	Global Positioning System
HE	high explosive
HEAB	high-explosive air bursting
HEF	high-explosive fragmentation
HEAT	high-explosive antitank
IAI	Israel Aircraft Industries
IMU	inertial measurement unit
INS	inertial navigation system
IR	infrared
LTTE	Liberation Tigers of Tamil Eelam
MANPADS	man-portable air defense systems
MBDA	a pan-European defense company
MBT LAW	main battle tank and light armored weapons
MFCS	multifire control system
MOA	minute of arc
MTCR	Missile Technology Control Regime
ODAM	Optically Designated Attack Munition
OICW	objective individual combat weapon
PAL	permissive action link
PDA	personal digital assistant
PGMM	precision-guided mortar munition
PIRA	Provisional Irish Republican Army
PLGR+GLS	personal lightweight GPS receiver plus gun-laying system

RAND-MIPT	RAND–National Memorial Institute for the Prevention of Terrorism
RF	radio frequency
RFGM	radio frequency guided munition
RPG	rocket-propelled grenade
rpm	rounds per minute
SAASM	Selective Availability Anti-Spoofing Module
TA/FC	target acquisition/fire control
TRAP	Telepresent Rapid Aiming Platform
VIP	very important person

CHAPTER ONE

Introduction

While considerable attention is being directed to potential terrorist use of unconventional weapons such as chemical, biological, radiological, and nuclear (CBRN) weapons, relatively little attention has been directed to potential terrorist use of advanced conventional weapons. The November 2002 attacks in Mombasa, Kenya, using Russian-made man-portable air defense systems (MANPADS) against an Israeli airliner, attributed by some to al Qaeda, demonstrated that some terrorists are willing and interested in using relatively unfamiliar, advanced weaponry.[1]

Our use of the term *advanced conventional weapons* is inclusive and broad: any new or unusual conventional weaponry developed for ordinary military forces. This is essentially the definition used by the U.S. Department of State, which describes advanced conventional weapons as "modern, sophisticated munitions designed for conventional warfare."[2] Such weaponry seems a priori likely to be particularly threatening in the hands of terrorists, as it is designed to do damage, while its sophistication might allow new, or at least unfamiliar, attacks. At the same time, the usual limitation of much weaponry to militaries also implies that some controls would be imposed. Of course, any controls on these conventional weapons would be less burdensome than those imposed upon nuclear, chemical, or biological weapons. This is still quite unlike the case of systems developed in the wider, commercial

---

[1] Caffera (2003), p. 13; Bayles (2003).

[2] U.S. Department of State (undated).

marketplace—where any control is unusual. Consequently, limiting the potential terrorist use of advanced conventional weaponry appears to be both important and possible, and therefore worth investigating.

Research indicates that most terrorist organizations are operationally conservative, favoring familiar weapons such as "the gun and the bomb" in carrying out operations.[3] This is not entirely surprising, as organizations' decisions to pursue new technologies and weapons will be driven by their assessment of the costs and benefits involved in doing so.[4] For many operations, familiar weapons and tactics are more than sufficient for groups to achieve the outcomes they desire and, as a result, there may be little reason for these groups to pursue them. Still, the Mombasa attacks and those on September 11, 2001, demonstrated that at least al Qaeda understood the potentially devastating and dramatic impact of unfamiliar and innovative attacks. A number of other advanced conventional weapons might be similarly attractive to terrorist groups, given their potential for devastating and dramatic impacts—outcomes that may be seen as valuable in terrorists' cost-benefit judgments. This book identifies two key characteristics of advanced weapons that shape other elements of terrorists' calculus about the value of pursuing these systems:

1. *Ease of use.* Many advanced systems have greatly reduced requirements for user training and expertise to operate the weapons. These ease-of-use qualities are usually leveraged through the integration of complex electronics, reducing both the risks and the "learning costs" associated with utilizing a new weapon rather than relying on familiar weapons and tactics—and therefore making them more attractive.

2. *Capability to overwhelm or circumvent security forces and their countermeasures.* Some advanced conventional weapons can

---

[3]   Hoffman (2000).

[4]   In the area of costs and benefits, we include not only financial costs of acquiring or making a new weapon, but also the time required for the group to learn how to use it, the risks (operational, security, and others) that the weapon will not perform as expected, risking operational success or the safety of operatives, and so on. A more complete discussion of these issues is included below in Chapter Five.

provide a terrorist group with firepower or destructive capabilities that overmatch current security forces or defenses. Even more worrisome are new weapons that would enable entirely new, and potentially unexpected, types of attacks. Such new attacks can circumvent existing security plans entirely, creating vulnerabilities that U.S. and other security agencies have not considered. The ability of these weapons to overwhelm current security measures shapes the benefit side of the terrorists' assessment of these weapons.

Study Approach

Given this context, two basic questions motivate this research effort:

- What difference would it make if terrorists could use advanced conventional weapons in their attacks?
- What could the United States do to reduce this threat?

Answering these questions requires a multistep analytic process. The process starts with an overview of the advanced conventional weapons that are currently in development around the world. It then extensively describes which of those systems appear to be most dangerous. This analysis focuses primarily on systems still in development, as opposed to advanced weapon systems already fielded, for two reasons. First, existing advanced conventional weapons, with the exception of MANPADS, do not appear to be attractive to terrorists because to date, we find no evidence that terrorists have attempted to use them. Presumably, the existing, perceived balance of costs and benefits does not seem attractive to terrorist groups. Second, imposing additional controls on a weapon is likely to be much easier before the equipment has been fielded—and arguably is feasible only then. Nonetheless, our discussion does address several types of existing advanced conventional weapons that could, under certain circumstances, be attractive for terrorist use, even though they have not yet been used by terrorists. This discussion is presented in Chapter Two.

The second step in the study was to assess the potential utility and attractiveness of these weapons to terrorists. This assessment, which is presented in Chapter Three, included two elements:

1. The assessment first investigated potential terrorist attack scenarios and the value of new weapons in those scenarios to explore the benefits of particular weapons to terrorists. Some of these attacks can be described briefly, since they are simple variations on current terrorist capabilities—for example, assassinating a political figure by using an improved sniper rifle. Others are unfamiliar and therefore require a longer description.

2. The assessment then examined previous terrorist use of existing versions of the most important, advanced conventional weapons that have been found in terrorist arsenals. The rationale is that, if a terrorist group is already familiar with a weapon class, such as mortars, it would be more likely to appreciate and thus attempt to exploit new, advanced capabilities within that class.

Third, the project team considered, in some depth, the potential for controls of different weapons. The analysis considers both procedural controls, such as those established internationally under the Wassenaar Arrangement,[5] and technical controls that might limit the functionality of a weapon that had been diverted from legitimate sources. Technical controls would exploit the growing use of electronics in most new weapons. Such limitations seem most plausible when the weapon system inherently relies upon a satellite navigation system —most commonly, the Global Positioning System (GPS),[6] but potentially the similar Russian or European systems. In practical cases, internal technical controls must be designed from the start of a system's research, development, test, and evaluation cycle; otherwise, they may

---

[5] Wassenaar Arrangement on Export Controls for Conventional Arms and Dual-Use Goods and Technologies (1998).

[6] Since GPS is the only widely available satellite navigation system today, the examples and discussion in this book all refer to GPS. The arguments we make involving GPS would hold for any similar satellite navigation system, however.

either be easy to circumvent or entail a very expensive redesign. This discussion of controls is presented in Chapter Four.

For the final step in the analysis, the research team examined whether plausible procedural and technical control regimes would have a large effect on the perceptions of terrorist groups. The study addressed this question through interviews with expert analysts who have studied different terrorist groups, asking whether the limitations that seemed possible would change the perception of the weapon's utility to the groups they study. Ideally, a control regime would make the controlled system thoroughly unattractive to terrorists. Combined with the direct limitations of the controls themselves, this reduced attractiveness, we assert, would greatly limit the threat from terrorists' use of such weapons. This discussion is presented in Chapter Five.

To conclude this monograph, Chapter Six draws implications from the analysis, identifying what measures seem available to the United States. The discussion ends with suggestions for the role that the U.S. Department of Homeland Security could play in proposing and assisting in the implementation of such measures.

CHAPTER TWO

What Types of Advanced Military Weapons Could Become Available to Terrorists?

Although most modern military weapons are more powerful than similar antecedents, many may be implausible for terrorist use. For example, there is little likelihood that a terrorist group would attempt to acquire and use a main battle tank or a jet fighter. Similarly, a large warship would be of little practical value to terrorists. As a result, the types of military weapons on which this book focuses are relatively small, person- or light-vehicle transportable systems that would be relatively easy to conceal and use. Most are systems that one person or a small group could employ. The research team sought to identify weapons with these characteristics and with steadily advancing capabilities that would be potentially attractive to terrorist organizations.

That screening effort resulted in the following list of weapons and devices for analysis:

- *Advanced small arms.* Rifles, pistols, and other individual weapons are easily carried by one person. Trends include better accuracy, greater range, and improved penetration of targets.
- *Mortar systems.* Mortars provide the ability to hurl an explosive projectile high into the air against targets several kilometers away. In recent years, mortars have gained precision strike capability through the integration of computer aids, GPS locators, and guided munitions.
- *Sniper systems.* Precision small arms, usually at or above 0.50 caliber and with new sighting and other aids, allow people lacking extensive sniper training to be able to engage targets at greater distances.

- *Antitank guided weapons (ATGW).* Wire-guided antitank missiles were the first examples of these systems. Available since the early 1970s, these weapons' range, accuracy, and ease-of-use have increased as their guidance systems have evolved.
- *Man-portable antiarmor weapons.* Epitomized by the rocket-propelled grenade (RPG) widely used in Iraq, this class of weapon has been gaining greater accuracy, improved range, and more warhead options.
- *Limpet mines.* Underwater explosive systems are designed to be attached to the hull of a ship by a scuba diver. Sizes range from 5 kilograms to several hundred kilograms. In use for almost 100 years, today's limpet mines include sophisticated antitamper devices to hinder their removal.
- *Advanced land mines.* Whereas in the past, land mines were "dumb" systems that were buried in the ground, today's advanced mines—usually emplaced above ground—have sensors and can attack targets from several tens of meters distance.
- *Night-vision devices.* Although not a weapon per se, modern night-vision devices provide significant tactical advantages to small units operating in darkness.

Additional detail on these systems and their evolving capabilities is provided in the following sections.[1]

Advanced Small Arms

A new generation of infantry small arms is presently being developed and fielded. These weapons include

- entirely new designs, such as assault weapons that launch computer-controlled smart grenades and 100 percent electronic jam-free guns that feature extremely fast rates of fire

---

[1]   It should be noted that, because the nature of the weapon technologies and their relevance in the later analysis (as discussed in subsequent chapters) differed, varied levels of detail are provided on each system, with more on the more relevant ones.

- new generations of existing rifles and light machine guns with improved modularity and sighting aids as well as low failure rates and extensive use of lightweight materials
- improved armor-piercing ammunition with limited penetration and high frangibility in soft targets for more lethal wounds.

The technological advances (e.g., in sensors, smart ammo, electronic firing, aim correction) currently seen in the designs of these weapons are leading to small arms that will be more formidable than ever before because they will allow for new tactics, will be more lethal, and will be easier to use. The computer aids for some of these systems include, for example, the capability to employ a laser rangefinder and environmental sensors to determine the optimal aim point and fuze setting for an airbursting grenade. A built-in rangefinder can determine the precise distance to a target, be it personnel or a wall, window, or building corner, while pressure and environmental sensors allow a ballistics processor to correct for the grenade's trajectory. The weapon's computer can then calculate and set the grenade fuze to produce airburst on, above, or beyond the aim point, while also indicating the appropriate aim correction to the operator through an in-sight display. Complementing the advanced sighting systems that are being built into new weapons or added to existing ones is more effective ammunition. Armor-piercing rifle ammunition that can penetrate most of today's personal body armor is widely available, and the trend toward even more penetrating ammunition, with the ability to more easily create untreatable, lethal wounds in soft targets compared with older same-caliber bullets, is well established.

These advancements could put guard personnel at important facilities, as well as police and quick-response forces reacting to a threat, at considerable risk for a number of reasons. For example, their body armor may be rendered obsolete by armor-piercing ammunition, or they may not be trained to protect themselves against an overhead, around-the-corner, or in-room attack from something like an airburst grenade.

Technological Advance:
Airburst Assault Weapons with Smart Ammo

Three weapons that use revolutionary airbursting smart ammunition are currently being developed:

- the Australian advanced individual combat weapon (AICW)
- the U.S. Army's XM25 airburst assault weapon
- the U.S. Army's XM307.

The features of the 14 lb semiautomatic XM25 airburst assault weapon (see Figure 2.1), which is currently being field-tested by the U.S. Army and is scheduled for fielding by 2008,[2] are representative of what will generally be available in this class of weapons.

Developed by Alliant Techsystems (ATK) and a spinoff of the abandoned XM29 objective individual combat weapon (OICW),[3] the XM25's features include

- the ability to use numerous types of magazine-loaded, low-velocity 25mm munitions, including thermobaric, flechette (anti-personnel), training, high-explosive airbursting (HEAB), door-breaching, armor-piercing, and nonlethal varieties[4]
- a full-solution target acquisition/fire control (TA/FC) system (known as the XM104), with integrated laser rangefinder, digital compass, ballistic computer, environmental sensors, and day and

---

[2] As of April 2005, six prototypes had been delivered to the U.S. Army for field-testing. U.S. Army project manager LTC Matthew Clarke has remarked that the "initial field tests are very promising." In one of the field tests, a grenade was launched through a small window at 170 meters downrange and was detonated inside a virtual room. See "XM25 Prototype in Testing" (2005) and "Army Will Boost Supply of Small Cal Ammo, Weapons" (2004). See also "ATK XM25 Grenade Launcher for Future Industry: Will It Fly?" (2006), and "XM25mm Airburst WeaponSystem" (2005).

[3] The XM29 included a semiautomatic 20mm multiple grenade launcher and an assault rifle but was abandoned due to a number of design flaws. See "ATK XM25 Grenade Launcher" (2006). See also "XM25 Individual Airburst Weapon System" (2005), and ATK (2005).

[4] See "XM25 Individual Airburst Weapon System" (2005). See also ATK (2005).

Figure 2.1
A Soldier Aims an XM25

SOURCE: U.S. Army (2005), p. 234.

night (thermal) sights with 500-meter ranges[5] (see Figure 2.2 for a diagram of the XM104 including components)

- maximum ranges of 300 meters against point targets and 500–700 meters against area targets as well as the ability to defeat defiladed targets[6]
- maximum ranges of 300 meters against point targets and 500–700 meters against area targets as well as the ability to defeat defiladed targets[7]
- four different modes for the "smart shells" with a computer-controlled fuze[8]:
  - airburst, in which the shell is optimized to spray incapacitating (wounding or killing) fragments in a roughly six-meter radius from the exploding round. Thus if enemy troops are

---

[5] See "XM25 Individual Airburst Weapon System" (2005).

[6] See ATK (2005).

[7] ATK (2005).

[8] See Murdoc (2004). It should be noted that these descriptions also apply to the 25mm airburst munitions in the crew-served XM307 being developed by General Dynamics because, in May 2003, the decision was made to combine the ammunition development efforts for the two weapons. See also Donovan (2003).

Figure 2.2
Components of the XM104 TA/FC for Use with the XM25

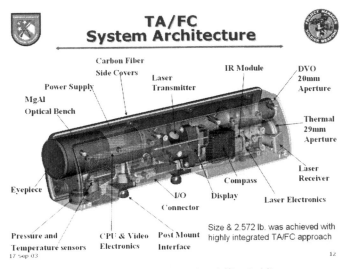

SOURCE: Courtesy of *Program Executive Office Soldier.*

seen moving near trees or buildings at a long distance (over 500 meters), the weapon has a good chance of getting them with one shot. M-16s are not very accurate at that range, and the enemy troops will dive for cover as soon as M-16 bullets hit around them. With smart shells, you get one (or a few) accurate shots and the element of surprise."[9]

— point detonation, in which the grenade detonates on contact
— point detonation delay, in which the grenade detonates after it has passed through a window, door, or thin wall
— window, in which the grenade detonates after it has passed the aiming point, whether that be a window or door frame or corner of a building.

It is mainly the combination of the TA/FC with the computer-controlled grenades that makes the XM25 different from any other individual combat weapon currently deployed. Figure 2.3 illustrates

---

[9]   See Murdoc (2004).

Figure 2.3
XM104 Sighting System and Ballistics Computer in Action

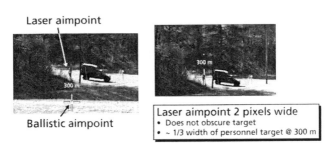

SOURCE: Courtesy of *Program Executive Office Soldier*.

the simplicity with which the TA/FC can be used to set the fuze of the grenades and to correct for their ballistic trajectory given the target's range and the environmental conditions involved in the shot. Basically, the operator need only follow two steps in order to hit the target.

In addition to the ease of use of the XM25, the publicly available and increasingly popular *America's Army* computer game, which will accurately simulate the weapon and its laser sight years before its fielding, will allow for virtual practice with the weapon (see Figures 2.4 and 2.5 for images of the weapon and its sight in the game). The fact that the computer game is intended for training can be seen in the following quote by Bill Davis, the team leader for future applications for the game: "The real key feature is the modeling of the fire control screen. It will give people a chance to try out tactics, techniques, and procedures in small unit settings."[10]

Because of the precise control over airbursting provided by the TA/FC, the XM25 could change the game between attackers and defenders who are not prepared for grenade detonations from above or even inside a sheltered position and from ranges greater than what is possible with a traditional launcher like the M203. This possibility has not been lost on the reviewers, developers, and testers of

---

[10] Tiron (2004). It should be noted that there are civilian and military versions of this game and that the civilian version may differ in that it would not use classified weapon performance data.

Figure 2.4
Accurately Modeled XM25 in *America's Army*

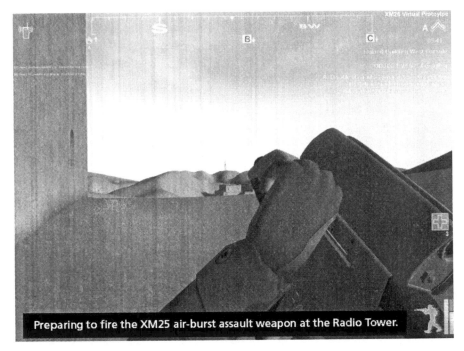

Preparing to fire the XM25 air-burst assault weapon at the Radio Tower.

SOURCE: U.S. Army (undated).

the weapon, who have called it variously "a great leap forward for individual fire support," "revolutionary," and a "clear differentiator on the battlefield."[11] Whether or not games like *America's Army* will spark an interest in the weapon in the United States' enemies and allow them to practice using it remains to be seen.

As mentioned, the XM25 is only one of three weapons being developed that use airbursting smart ammunition. The XM307 and the AICW are examples of other weapons with similar features currently under development for the U.S. Army and the Australian Army,

---

[11] Murdoc (2004); ATK (2005); and LTC Matthew Clarke, the U.S. Army project manager for individual weapons, quoted in Gizmag.com (2005).

Figure 2.5
XM25's Fire Control Screen Simulated in *America's Army*

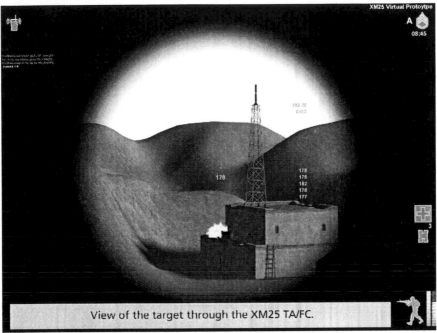

View of the target through the XM25 TA/FC.

SOURCE: U.S. Army (undated).
NOTE: The future applications team leader thinks of this as the key to providing useful training through the game.

respectively. The XM307[12] (see Figure 2.6) is a replacement for a heavy machine gun being developed by General Dynamics that fires the same ammunition as the XM25 does and, also like the XM25, includes a full-solution fire control system with a laser rangefinder and day and night sights. The XM307 is scheduled for fielding by the end of fiscal year (FY) 2008.[13] The AICW is being developed by Metal Storm and

---

[12] General Dynamics' XM307 has some advantages (e.g., a 2,000-meter range) and disadvantages (e.g., it weighs 50 pounds and is meant to be portable by only two people). Another advantage of the XM307 is that it can be converted into a lightweight 0.50-caliber machine gun known as the "XM312 in minutes." See "XM307 25mm Airbursting Weapon System" (2005) and Kennedy (2005).

[13] See General Dynamics (2005) and Kennedy (2005).

Figure 2.6
The XM307 Has Many Similarities to the XM25

SOURCE: Courtesy of *Program Executive Office Soldier.*
NOTE: The XM307 is intended as a replacement for the heavy machine gun. One of its similarities to the XM25 is its smart airbursting shells.

Tenix as a concept technology demonstrator (CTD) for evaluation by the Australian Army and was successfully test fired on August 31, 2005 (see Figure 2.7).[14] The development efforts are being led by the Australian Defence Science and Technology Organisation (DSTO) in a three-year research and development program expected to end in late 2005. The design of the AICW is something of a cross between the XM25 and the scrapped American XM29 OICW. Like the XM29, it will fire grenades and have an assault rifle component. Like the

---

[14] See Davide (2005).

**Figure 2.7
A Soldier Test Fires the AICW**

SOURCE: Courtesy of Metal Storm.

XM25, it will feature smart airbursting ammunition and include day and night sights, a fire control system, and a laser rangefinder.

**Technological Advance:
Metal Storm's 100 Percent Electronic Firing Mechanism**

Although this airbursting smart ammunition represents a revolutionary change in the tactics that are possible with small arms, Metal Storm's fully electronic firing mechanism, in which the projectiles are the only moving parts, represents a revolutionary improvement in the speed, reliability, flexibility, and lethality with which they can be used. In the

Metal Storm technology, projectiles are stacked one after another in a barrel instead of in a magazine. Each projectile, whether it be a handgun bullet or a 40mm airbursting round, has its own propellant load and can be electronically ignited. A diagram of the AICW is shown in Figure 2.8. Two capabilities offered by the Metal Storm technology are that there are no parts to jam and that weapons can be designed for area denial, special operations, personal protection, combat, sniping, and other uses depending on how barrels, each one of which can hold a different type of ammunition, are combined (see Figure 2.9 for a handgun concept that features four barrels, two with lethal and two with nonlethal ammunition). Other advantages include the ability to fire extremely quickly and to fire multiple shots in bursts before feeling the weapon's recoil,[15] increasing the likelihood of penetrating armor and killing a target with one shot or one burst. This first-shot kill capability may be especially important for sniper rifles developed with the technology.[16] Clearly, if only one side in a confrontation has weapons that offer these advantages, it may give that side an edge. Although the Metal Storm electronic technology enhances the offensive capabilities of weapons that employ it, it also allows for use controls that are not possible in weapons with mechanical movement. For example, the prototype VLe handgun utilizes a transponder ring worn by the user

---

[15] For example, the VLe electronic handgun prototype fired very short bursts either at an effective rate of 45,000 rounds per minute (rpm) when the option of two rounds per trigger pull was selected (called a *double tap*) or 60,000 rpm with the three-rounds-per-trigger-pull option (a *triple tap*). According to Metal Storm, these shooting speeds make it look as if only one shot has been fired when, in reality, "three impacts appear virtually simultaneously on the target." In later demonstrations for the Defense Advanced Research Projects Agency (DARPA), it fired longer bursts at similar rates of fire, emptying a small magazine. In all cases though, these rates of fire cannot be maintained for any significant time; rather, the advantage is that all the rounds are tightly bunched in space. For comparison, a fully automatic Glock 18c pistol fires an average of 500 rpm. See Metal Storm (2002) and Generation Airsoft (undated).

[16] According to Metal Storm (2004b), four shots can be fired in one burst from a 0.308 caliber rifle using its technology, allowing for "compounded kinetic energy from a smaller and lighter weapon that nearly matches the lethality and penetrating power of the model 82A1 .50 caliber Sniper Rifle."

Figure 2.8
Diagram of the AICW

SOURCE: Courtesy of Metal Storm.

to activate the gun.[17] Also, the operating system used by Metal Storm weapons will allow for such future controls as fingerprint identification.[18]

Understanding the potentially game-changing Metal Storm technology and how best to control it is important because this technology is finding its way into a number of developmental weapons that have been successfully test fired, including the VLe handgun, the grenade launcher on the Australian AICW, an area denial weapon system that features 40mm projectiles stacked in four barrels, and a four-barrel 40mm system mounted on a Talon Unmanned Ground Vehicle.[19] Metal Storm technology is even being considered for use in U.S. homeland security. For example, as of August 18, 2005, the U.S. Department of Energy has been negotiating a contract with Metal Storm for research and development of a 40mm short-range neutralization system.[20] It makes sense that care should be taken so that this technology more ensures U.S. security than threatens it.

---

[17] See Metal Storm (2004a).

[18] See Metal Storm (2004c).

[19] See Metal Storm (undated).

[20] See Davide (2005).

Figure 2.9
A Four-Barreled Concept Handgun

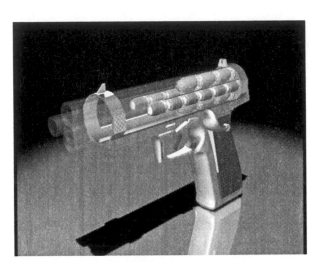

SOURCE: Courtesy of Metal Storm.
NOTE: This four-barreled handgun would shoot two
types of ammunition and use Metal Storm's electronic
firing technology.

Mortar Systems

Mortars have long been regarded as cheap, lightweight, short-range artillery. Mortars are generally small weapons, certainly much smaller and more compact than artillery pieces. Modern mortars are also relatively lightweight. For example, the U.S. M224 60mm mortar weighs 18 to 45 pounds, depending on the configuration. Its ammunition is four pounds per projectile. The M252 81mm mortar is 93 pounds, with 15-pound ammunition. Finally, the M120 120mm mortar is 320 pounds, firing a 33-pound projectile. Artillery pieces, in comparison, weigh far more, 9,000 to 15,000 pounds being typical for 155mm howitzers.[21]

---

[21] See "Mortars" (1998).

When firing traditional, unguided projectiles, mortars are used as barrage weapons that cover an area with rapid fire. In recent years, however, the trend has been toward the creation of precision munitions for mortars, initially infrared-guided projectiles, and subsequently laser-guided rounds and fiber-optic–guided munitions.[22] Simultaneous with the improved accuracy of precision munitions, the range of mortars is also increasing. Whereas the Soviet-era 120mm mortar had a maximum range of about 5.7 km, new modern 120mm mortars are capable of ranges well over 10 km—and the trend toward increased range is continuing. Additionally, the variety of munitions that can be fired by mortars is also increasing. In the past, mortars were limited to firing high-explosive (HE) and smoke rounds. Today, submunitions-filled projectiles that scatter small bomblets around the target area as well as precision munitions, such as those noted above, are now available. Modern mortar systems are also relatively portable and require few individuals to operate.

Advances in military mortar system technology have been fueled by a desire to give the battalion or company commander a piece of "hip-pocket artillery" that allows for "shooting and scooting" while delivering a precise and lethal strike. Being indirect-fire weapons that do not require line of sight to the target, mortars allow their operators to shoot from a defiladed position[23] and improve the prospects of escaping after the attack with the attackers' anonymity preserved.[24] Advances that make it even easier to shoot and get away include mortar rounds capable of doubling current ranges (from 7 to 15 km for the largest of the tactical mortars[25]) through the use of wings or rocket motors and components designed for quick deployment that are both

---

[22] Laser-guided rounds require that the target be designated by an observer with a laser target illumination device.

[23] Current man-portable mortar systems are evolutions of a design from World War I, which saw the need for weapons with high attack angles that could be fired into enemy trenches from protective cover. See "Mortar" (undated).

[24] The attackers in a majority of "terrorist" mortar attacks are unidentified (see Chapter Three on terrorist experience with mortars).

[25] Mortars between 60mm and 120mm in size are often described as "tactical" mortars.

smaller and lighter. Precision has greatly increased due to homing or man-in-the-loop terminal guidance systems[26] coupled with rounds that are capable of nonballistic flight.[27] Also, both new systems and systems undergoing midlife improvement are being equipped with computer aids that use GPS, meteorological, and topological data for setting the fire angle and azimuth, simplifying the use and potentially increasing the effectiveness of even ordinary "dumb" rounds. Full GPS-only solutions that eliminate the need for manual terminal guidance will also eventually be available.[28] These aids not only increase precision, but also greatly reduce artillery training requirements and the time needed to prepare to fire. Finally, the lethality of mortar rounds has been increased through the development of submunitions, high-explosive fragmentation (HEF) warheads, and other payloads.[29]

Technological Advance: Gliding and Rocketing to Longer Range

Extended-range 120mm mortars have used two methods to increase their maximum range to more than 15 km: rocket propulsion after launch and gliding using deployable wings. Only one extended-range system uses a rocket motor: Talley's XM984 120mm extended-range

---

[26] Terminal guidance systems are systems that help guide the mortar to its target once it is already past the highest point of its flight. Such systems consist of some way to steer the mortar round (e.g., control fins or side thrusters) and some sort of sensor on the warhead (e.g., an infrared [IR], radio frequency [RF], or laser-seeking sensor or a video camera). Terminal guidance can involve a man in the loop when an operator is used to point a laser at the target or to steer the mortar using the video feedback. Terminal guidance can also be a homing system in which the mortar simply tracks a known signature or frequency, and no man in the loop is required. More information on terminal guidance can be found later in this chapter.

[27] A conventional mortar is a ballistic weapon propelled out of a tube. Its trajectory is affected mainly by muzzle velocity (the speed with which it leaves the tube), gravity, air resistance, and other environmental factors. Some advanced mortars are able to modify this trajectory through the use of rocket engines, wings, control fins, or side thrusters. More information on nonballistic trajectories can be found later in this chapter.

[28] The ability to aim a mortar simply by entering the GPS coordinates of the tube and of the target will be a feature on systems currently in development such as the Israeli Fireball (read more below).

[29] Some of these new payloads and warheads increase the area that can be covered, while others increase the types of armor that can be penetrated.

dual-purpose improved conventional munition (ER-DPICM), which was tested by Talley in 2001 and has a range of 11 km.[30] The XM984 is designed to carry a variety of payloads and has a longer range, but it *does not* include precision terminal guidance (see Figure 2.10). An electronic fuze is used to set the times of both rocket ignition and payload release. The design for such a fuze was in development (also by Talley) as recently as 2002, meaning that opportunities for controlling its design may still exist (see Figure 2.11). For example, during a test of three prototype rounds, the first two had rocket ignition delays set to six seconds and achieved 7.8 km ranges, while the third round reached 8.7 km with a 12-second delay.[31]

Unlike the XM984, all of the precision systems have opted for the deployable-wings design to allow final course corrections. These systems, including their fielding dates, estimated ranges, manufacturers, and countries of origin, are as follows:

- ATK's 120mm M395 precision-guided mortar munition (PGMM)[32]
  - scheduled for fielding in 2006
  - maximum ranges of 12 km (threshold) to 15 km (objective), according to the requirements of the development program
  - ATK is a U.S.-based advanced weapon and space system company
- Israel Aircraft Industries (IAI) 120/121mm Fireball[33]
  - currently in the advanced development stage (all critical components have been fire tested)
  - maximum range of 15 km

---

[30] Talley Defense Systems is a U.S.-based developer and producer of propellant-based products. See "Talley Defense Systems" (undated). The XM984 is also referred to in some sources as an extended-range mortar cartridge (ERMC). See Pascua (2002). See also "Mortar Systems" (undated).

[31] See Yoo (2002).

[32] See "M395 Precision-Guided Mortar Munition" (2005).

[33] IAI (undated, 2002).

Figure 2.10
**XM984 Uses a Rocket Motor to Reach Up to
11 km**

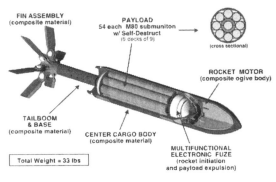

SOURCE: Pascua (2002).

Figure 2.11
**Israel Aircraft Industries' Fireball Has Four Deployable Wings as Do the
PGMM and Armement Léger à Fibre Optique (ALFO)**

SOURCE: Courtesy of *Defense Update*.

- MBDA's 120mm ALFO Lightweight Fiber Optic Weapon[34]
  - entered the development phase in 2002; discontinued by 2006
  - maximum range of 20 km.

As an example, Figure 2.12 shows the components of the PGMM and its flight profile. Although the complexity of the round is increasing, with guidance, wings, and the like, it remains small enough for a single person to easily drop it into the mortar tube. During flight, the wings deploy at apogee, and the mortar glides for an extended range.

Both the rocket-assisted and gliding 120mm mortars listed above effectively at least double the 7–8 km ranges of traditional rounds. Although those systems represent some of what is under development in terms of increasing range, this is not a comprehensive list (e.g., in the 1990s, the Chinese developed an extended-range 120mm mortar that could reach 12 km).[35]

Advances have been made in increasing the range of smaller mortars as well. For example, a proposed 81mm ALFO, which was based on the same design as its big brother, could have reached distances of 7 km, and an extended-range 81mm round developed by China can reach 8 km. Both significantly improve on the 5–6 km range of a typical mortar of this size from the 1980s.[36] When it comes to 60mm rounds, the South African Vektor M6 and the Chinese Type 90 have ranges of 6 and 5.7 km, respectively, effectively doubling the 2–3 km capabilities of traditional 60mm mortars. Both the Vektor and the Type 90 are already in production.[37]

---

[34] MBDA is a pan-European defense company that maintains a strong presence in the UK, Italy, and France. See MBDA (undated).

[35] The Chinese have not only developed extended-range mortars similar to those of Western countries but are also working on developing guided mortars. See "Type 35" (undated), "Type 64" (undated), and "Type 86" (undated).

[36] See MDBA (undated); see also "Type 35" (undated), "Type 64" (undated), and "Type 86" (undated).

[37] See "Anti-Armor Missiles" (2005). See also "Type 35" (undated), "Type 64" (undated), and "Type 86" (undated).

Figure 2.12
The Components and Flight Profile of ATK's PGMM

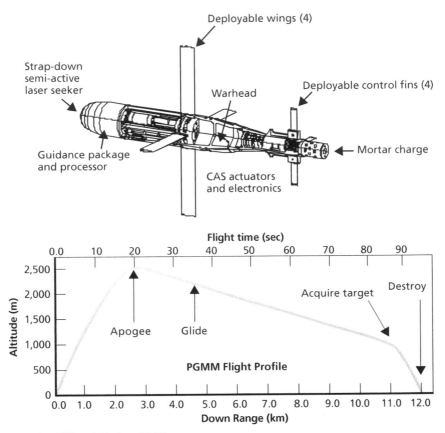

SOURCE: Cilli and Bischer (2000).

These extended-range systems allow attacks to be carried out from distances so great, especially in the case of the 120mm mortars, that no reasonable security perimeters could be formed to protect against them. Also, the area that would have to be searched to find the attackers would be sufficiently large to give them ample time to escape.

Technological Advance:
Nonballistic Flight Through Thrusters and Control Fins

The extended-range flight paths made possible through wings and rockets discussed above are nonballistic trajectories.[38] Other technologies, however, also allow for nonballistic trajectories. For example, additional fine steering in systems with terminal guidance has been implemented through the use of thrusters and control fins.

One precision 120mm system that uses trajectory-correction side thrusters controlled by guidance computers is the Saab Bofors STRIX, which has been in service with the Swedish Army since 1994.[39] It has a conventional 7 km range but uses terminal infrared homing and steers to its target at the end of its ballistic flight.[40] These systems are designed to attack wheeled and tracked vehicles with a characteristic thermal signature and thus are of limited utility. For example, they cannot distinguish one similar vehicle from another, making precision attacks on a particular car, such as a presidential limousine, and thus assassination through IR homing, unlikely.

Precision systems developed since the STRIX, including the PGMM, the Fireball, the proposed ALFO, and the Gran, use control fins instead of thrusters.[41] While the mortar is in flight, an inertial measurement unit (IMU) calculates its position. The control fins use information from the IMU and seekers on the warhead to guide the mortar first to an "air basket" and then to the target.[42]

---

[38] They are already nonballistic paths because their trajectories are different from what the mortar would follow were it affected only by muzzle velocity, gravity, air, and other environmental factors.

[39] If 1994 seems much earlier than the other development dates in this chapter, it is because the STRIX is indeed known as the "first really smart mortar bomb in service in the Western world" (Ripley and Biass, 1998).

[40] Saab (undated) and "STRIX Precision Guided 120mm Mortar Launched Weapon" (2004).

[41] The Gran is a 120mm system designed by Russian arms manufacturer KBP that was fielded in 2000. It has laser-homing terminal guidance but no extended range capability.

[42] An air basket is an approximate area in the air appropriate for delivery of the munition to the desired target. This description refers specifically to the PGMM. Other systems function similarly, however. The exception is the Gran, which flies a ballistic trajectory and then

The advantages offered by such thrusters and fins is that quick, one-shot defeat of both moving and stationary targets is now possible without attackers needing to aim at them with perfect ballistic trajectories. Of course, some sort of guidance is needed to take advantage of this maneuverability.

Technological Advance:
"Fire-and-Forget" IR- and RF-Homing Terminal Guidance

The earliest smart mortar, the STRIX, uses an imaging infrared sensor to discriminate among the true target and both decoys and burning fires. As described above, side thrusters on the mortar are used with IR-homing to guide it to the target.[43]

Advances in mortar technology have led to such fire-and-forget homing technologies being combined with man-in-the-loop laser designation (described below), resulting in dual-mode weapons. The PGMM, for example, has been envisioned as having an imaging infrared sensor in addition to the laser seekers to allow for a fire-and-forget mode.[44]

Other fire-and-forget advances just around the corner include the ability to home in to a target not only using infrared signatures but also using radio frequencies. An example is BAE's 81mm radio frequency guided munition (RFGM). The end goal of the RFGM development project, awarded by DARPA in January 2005, is to create an 81mm mortar that can use an RF seeker to find a radio or jamming device operating between 30 MHz and 3 GHz and then maneuver in flight to strike the target at the calculated location. The phase 1 effort is expected to last nine months and will include development of the antenna and receiver hardware as well as the signal processing software.[45]

---

fine-tunes its attack angle to the target like the STRIX. See "Precision Guided Mortar Munition" (undated) and "Gran 120mm Guided Mortar Bomb" (2004). See also Shipunov et al. (2003).

[43] See Saab (undated).

[44] See "STRIX Precision Guided 120mm Mortar Launched Weapon" (2004) and "Balad Airbase" (undated).

[45] See "BAE Systems Designs Precision Seeker for Mortar Rounds" (2005).

The potential advantages of such fire-and-forget systems over a standard mortar are limited by the need for the target signature. Nevertheless, because the mortar corrects its flight to engage a target, the attackers have a higher probability of hitting a target that has a known signature or frequency emission. Such terminal homing also reduces the need for conventional artillery training and the amount of setup time required to carry out an attack and the number of rounds that might otherwise be expended to hit the target.

Technological Advance:
Laser and Fiber-Optic Man-in-the-Loop Terminal Guidance
Two main types of man-in-the-loop terminal guidance systems for mortars have been fielded or are currently in development: One uses a laser to illuminate a target and a warhead equipped with semiactive laser seekers, while the other uses a video camera mounted in the warhead and a fiber-optic data link that allows an operator at the launcher to see and steer toward the target. These types of terminal guidance are used in conjunction with systems that use GPS and other types of data to get the mortar round into an "air basket" from which it can then use fins or thrusters to steer toward the target. They are being developed because they allow for much better precision than is possible with a simple gun-laying system (GLS, described below) and because they can hit targets that do not have the RF or IR signatures required by the fire-and-forget homing mortars (described above).[46] Man-in-the-loop guidance provides these systems with sufficient precision to attack relatively hard targets, such as vehicles or bunkers.

Laser-designated systems have been under development in many countries for a number of years and will be available for all sizes of tactical mortars, from 60mm to 120mm rounds. One hundred twenty–millimeter systems include the Gran (fielded in 2000) and the 120mm PGMM (under development and scheduled for fielding in 2006). The Fireball is another 120mm system that includes seeker technology,

---

[46] A GLS is a computer that uses GPS and other data (such as ballistics characteristics of the mortar rounds) to output the correct fire angle and azimuth of the mortar tube to hit a target.

which also fits into an 81mm mortar.[47] In addition to these three systems, a 60mm laser-homing mortar called the Optically Designated Attack Munition (ODAM), for which the development contract was awarded by DARPA in February 2005 and was scheduled to be completed in 24 months, is being designed by BAE Systems, with a live-fire test expected at the end of the contract.[48]

Although there are numerous laser-based systems, MBDA's ALFO, also described above, would have been the only fiber-optic–guided mortar being developed. It would have been available, however, in the broad array of man-portable sizes: from the 60mm micro-ALFO to 120mm configurations.[49] The ALFO system would have been controlled though touch-screen laptop computers or, in the case of the micro version, through a helmet-mounted display system.

Most of these systems have similarly impressive precision capabilities. The PGMM and the Fireball, for example, are being designed to hit targets with a 1m circular error probable (CEP).[50] The technical requirements document for the ODAM describes the expected CEP as being 4m for targets within 100m (and eventually 200m) of the mortar's ballistic trajectory. The Gran is described as having a hit probability of at least 90 percent.[51] All of these accuracies are sufficient to hit a small target.

There are two main effects of such precision. One is that, instead of using dozens of rounds while adjusting fire to reach a target, a one-shot defeat (i.e., the ability to kill a target with one shot) is now possible (see Figure 2.13). The other is that, within a small area, multiple targets can be attacked without changing tube elevation or projectile settings. For example, the Gran can launch mortars at individual targets within a 300-meter–radius area solely through use of the laser designator.[52]

---

[47] See IAI (2004).

[48] See BAE Systems (2005).

[49] See MBDA Missile Systems (2003, undated).

[50] See Bischer (1999); also see IAI (2004).

[51] See Shipunov et al. (2003).

[52] "Gran 120mm Guided Mortar Bomb" (2004).

Figure 2.13
One-Shot Defeat Is Possible for Some Precision Mortars

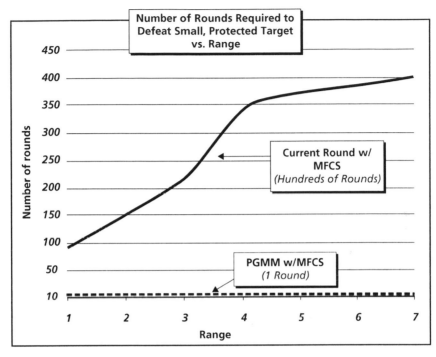

SOURCE: Bischer (1999).
RAND *MG510-2.13*

When it comes to defending against attacks, the most important difference between fire-and-forget IR or RF homing as described in the preceding section and man-in-the-loop laser-homing is the fact that the laser-homing technology requires an operator armed with a laser designator to be within 1 km of the target and have line of sight to it, thus creating a specific activity that may be observable within a reasonable perimeter. The fiber-optic system does not require a similar operator with a line-of-sight requirement, and it has the distinct advantage of allowing the attacker to be as far away or even farther from the target than the mortar launcher.

Technological Advance:
GPS-Based Computer Aids, Integration, and GPS-Only Guidance

The PGMM, Gran, Fireball, and ALFO precision systems described above make use of GPS to make mortar guidance easier and more accurate.[53] The coordinates of the shooter and the target are used to set the elevation of the tube and its azimuth (the direction to the target). In the case of the precision extended-range systems, the target's location is also passed to the mortar round, which uses the shooter's initial location and an IMU to calculate its position during flight and correct its trajectory to the target.[54]

Using GPS information, the mortar can be guided first to an air basket, which in the case of one system has a footprint on the ground of 500m by 500m,[55] from which an operator then steers it toward the target via fiber optics or guides it there with a laser designator.

Some mortar systems will be able to use GPS information alone to steer toward the target when something like a laser is not available. IAI's Fireball is described, for example, as being capable of GPS-only operation, which results in a hit accuracy that is greater than that of standard mortars but smaller than that of mortars using terminal guidance.[56] Such GPS-only guidance is already being tested in 155mm artillery rounds, providing an accuracy "better than 10-meters."[57]

Given this success in artillery, GPS and IMU components can obviously be built to withstand the much lower acceleration of a mortar round. And just as for artillery, this accuracy can still allow a single shot to destroy a target. In particular, for attacks on soft targets such as exposed personnel (one potential terrorist target) an accuracy

---

[53] See MBDA Missile Systems (2003) and IAI (2004).

[54] This is the case with the extended-range systems: PGMM, the Fireball, and ALFO systems. The Gran uses GPS and other data only for the initial fire preparation. See Shipunov et al. (2003).

[55] See Yoo (2002).

[56] See IAI (2004).

[57] Raytheon Company (2005).

of 10 meters is less than the lethal radius of a single mortar shell.[58] For now, the relatively high cost and size of the available GPS and IMU components prevent an easy integration into mortars. Additionally, the mortar round would almost certainly need to interact with another GPS shortly before launch, such as the GLS discussed below; this would provide to the mortar round a current position and timing information and, ideally, the positions of the current GPS constellation. With that information, a round could easily acquire the exacting P(Y) code of the GPS while in flight, enabling it to then achieve accuracies near 10 meters.[59]

GPS-only solutions provide worse accuracy than those with some form of terminal guidance simply because of the intrinsic errors of the GPS. When terminal guidance through homing or laser designation is used, however, the mortar, even with an imperfect calculation of its absolute location, can get close enough to "see" the target and thus discover both its relative position and the necessary attack angle (though at the expense and operational complexity of using a terminal seeker).

In addition to being part of many terminal-guided systems, GPS is used in many GLSs and midlife improvements to existing deployed mortars. An example of a GLS that is currently available is Rockwell Collins' personal lightweight GPS receiver plus GLS (PLGR+GLS). It uses a standard GPS receiver to tell the operator his or her location. Given this location and the target location, the GLS can calculate an accurate azimuth (i.e., the direction in which the tube should point) within three mils in less than five minutes (a circle is made up of 360 degrees or 64,000 mils).[60] Figures 2.14 and 2.15 show a PLGR+GLS

---

[58] "Ammunition for 81mm M29 and M29A1 Mortars" (2003) shows a bursting radius for the HE round of 17 meters. Larger 120mm mortars have a larger radius still.

[59] Some types of GPS/IMU combinations in use cannot arrive at satisfactory solutions given short times of flight. However, fast acquisition GPS receivers have been demonstrated on developmental systems that meet the timelines (if not cost and packaging requirements) for applications in guided mortar systems.

[60] See Ripley and Biass (1998).

Figure 2.14
Rockwell Collins' PLGR+GLS in Use

SOURCE: Courtesy of Rockwell Collins.

in use and its accompanying handheld GPS receiver, respectively. A simple GLS such as PLGR+GLS requires that the tube be adjusted manually according to the outputs that the GPS unit and the GLS provide to the operator.

There are, however, portable fire control systems that use GPS and include mechanical servos that adjust the mortar tube automatically. An example of such a system is the deployed South African Mechem multifire control system (MFCS), which can be placed on a pallet along with the mortar itself for rapid deployment on and off a carrier vehicle.

**Figure 2.15
A Handheld Personal Lightweight GPS
Receiver (PLGR)**

SOURCE: Courtesy of Rockwell Collins.

Figure 2.16 shows the MFCS being used with a 60mm mortar.[61] An example of a deployed system that has received many GPS-based benefits as part of a midlife improvement effort is the UK 81mm L-16. A GPS automatic GLS (AGLS) and a laser rangefinder with integrated

---

[61] "Infantry" (undated).

**Figure 2.16
The Mechem MFCS in Use with a
60mm Mortar**

SOURCE: Courtesy of Denel.

GPS were both added to the system in 2000.[62] Whether or not they are used in conjunction with terminal guidance, GPS aids increase shot accuracy and reduce the artillery training required for a successful attack. They also increase the speed with which attacks can be carried out by reducing the calculations that must be done to aim correctly. Without these aids, attackers attempting to use a mortar system would have to be familiar with using plotting boards and range tables.

**Technological Advance:
Lightweight Materials and Design Changes to Increase Speed**
A number of improvements in the way mortar systems are designed and in the materials from which they are made are increasing the speed with which they can be deployed and the ease of transport. Examples of design changes that make transporting and deploying a mortar system easier and faster are a base plate and carriage that allow for a 360-degree–firing azimuth without moving the plate (e.g., this is a component of the Gran) and the use of a pallet that includes the fire control system and the mortar components in one easy-to-deploy package (e.g., the Mechem MFCS and mortar can be deployed in this

---

[62] Shipunov et al. (2003) and Gander (2000).

manner).[63] An example of an advance in lightweight materials for mortars is the development of the carbon fiber composite barrel in the United States and South Africa. The importance of this advance is that it could make 120mm systems as easily transported by one soldier as an 81mm system is.[64] Regardless of whether the change is in the materials used for a component, the design of a component, or the grouping of components, any step that manufacturers take in making their weapons easier to carry and faster to deploy could make them more attractive to any attacker that would like to "shoot and scoot."

Technological Advance: Greater Penetration, Greater Area Coverage
New mortar warheads and submunitions are being developed for greater area coverage and to defeat more types of armor and other security measures (e.g., building-hardening). Mortars are being developed to accept numerous and different payloads. An example of a mortar that uses submunitions and can accept a wide variety of payloads is the XM984. It can drop 54 submunitions at a time specified by an electronic fuze and achieve both a 233 percent increase in area coverage and an 82 percent increase in antipersonnel effectiveness over a conventional mortar. Alternatively, it can accept six mines, a thermobaric warhead, a smoke bomb, and many other types of payloads.[65]

Conclusion
Many of the systems currently fielded or being fielded in the near future combine a number of these new technologies. The Gran, for example, combines a computer aid that gets mortar rounds into an air basket with a man-in-the-loop laser designation system that not only allows for precision kills but also for hitting various targets within a small area without adjusting the tube elevation. This precision is combined with an HEF warhead, components designed for fast deployment,[66] and the

---

[63] Shipunov (2003) and Gander (2000).

[64] See Yoo (2002).

[65] Pascua (2002).

[66] The estimated times involved in using this system are two to three minutes for computer deployment, 15 minutes for getting ready to fire after detecting a target, and then three min-

ease of use that a computer aid provides. Newer precision systems being fielded in the near future add such improvements as an increased maximum range of up to 20 km and wire guidance from the launch location. Many are being designed for launch from existing mortar tubes, meaning that a completely new system, tube and all, would not have to be acquired to take advantage of the advanced capabilities offered by new rounds.[67] Additionally, newer systems are being created that run the whole gamut of man-portable tactical sizes, from 60mm to 120mm, and with modular civil technologies to keep down costs and allow for mass production.[68] If volume production indeed becomes the norm, controlling the proliferation of advanced mortar systems could become both more difficult and more pertinent for security forces.

Although mortar systems are becoming more portable and are being geared toward urban warfare,[69] current countermortar systems, including acoustic arrays and counterbattery radars to detect incoming rounds, are designed for field combat (as opposed to defense in a civilian environment), in which artillery can fire and suppress further mortar launches. Such fire is implausible outside of a war zone and would likely fail to counter the first-shot capability of these systems even if it were plausible. High-energy lasers, envisioned to intercept and destroy or disable incoming mortar rounds, are not near maturity and would also seem implausible for domestic use if they were mature because of their high costs and potential for injuring civilians.

---

utes for actually getting the mortar tube into and out of action. The computer is made to be carried easily in three backpacks, and the mortar, though weighing approximately 400 kg, can be carried by a group of people or in a transport vehicle (see Shipunov et al. [2003]).

[67] See MBDA Missile Systems (undated).

[68] For example, the ALFO uses off-the-shelf fiber optics, and its sensors can be swapped out as technology improves.

[69] MBDA Missile Systems (undated).

Sniper Systems

The marksmanship chapter in a 1990s U.S. Army field manual on sniper training opens with the following statement:

> Sniper marksmanship is an extension of basic rifle marksmanship and focuses on the techniques needed to engage targets at extended ranges. To successfully engage targets at increased distances, the sniper team must be proficient in marksmanship fundamentals and advanced marksmanship skills. Examples of these skills are determining the effects of weather conditions on ballistics, holding off for elevation and windage, engaging moving targets, using and adjusting scopes, and zeroing procedures. Marksmanship skills should be practiced often.[70]

In recent years, considerable advances have been made in sniper technology that have changed the requirements for effective use of sniper tactics. These advances include

- ballistic computers that correct for factors affecting the bullet's trajectory such as wind and temperature
- platforms that stabilize a rifle and allow it to be operated by remote control
- improved scopes and reticules to assist sighting
- night-vision devices that allow for long-range shooting in the dark or other conditions of degraded visibility
- environmental sensors that reduce the need for a sniper to have detailed knowledge of the environment.

Technology has made it so that advanced rifle marksmanship skills are no longer necessary in sniping. The effect of these advances has been to

- reduce the need for a sniper to know ballistics and have advanced shooting skills

---

[70] U.S. Department of the Army (1994).

- make the sniper rifles more accurate
- increase the effective range of rifles from a few hundred meters to a mile or more (over 1,500 meters)
- make target acquisition faster and possible under a wider range of conditions
- make the impact of the ammunition more devastating to the target.

New systems that can be added to sniper rifles extend the traditional range (100m to 1 km) at which they can successfully hit a target, potentially enabling a first-shot kill at 2 km or more. Many of these components are designed to work together and produce clear instructions for the sniper on how to modify his or her aim to engage a target, thus reducing error, the amount of training and knowledge required, and the time needed to set up a shot. Because of the demand for firearms technology by established sporting communities, a number of these technologies are widely available on the open market.[71] Although a fairly well-trained shooter can better take advantage of the potential of these enhancements, these improvements can provide even a novice much greater ability to hit a target at greater range than would normally be the case for an untrained shooter. This has made it possible for the marksman who is not trained in long-range hunting to have (potentially) an easier time hitting a target at distances over 1 km.

The impact of a technology should not be judged merely by its existence, but also by its accessibility and the availability of information on how to use it effectively. Unlike other weapon systems covered in this monograph, most types of equipment for long-range hunting and tactical shooting are available legally and inexpensively to civilians; information on their use can be easily gathered through instruction manuals, Internet chat groups, online demos and lessons, video games, training courses, and live-fire demonstrations. For example, many different manufacturers of equipment provide interactive demos

---

[71] In addition to these add-ons that are available to the public, there are military-use-only sniping technologies being deployed, such as remote-aiming platforms that allow a target to be a hit without requiring even the most basic marksmanship skills.

that contain step-by-step instructions for using their product, down-loadable manuals, answers to frequently asked questions (FAQs), and courses on everything from marksmanship fundamentals to advanced long-range ballistics.[72] Chat groups run by both hobbyists and profes-sionals provide advice on very specific topics (e.g., the effect of spin drift on a long-range shot).[73] Finally, as was mentioned above, the would-be sniper can learn about and practice shooting using Web-based or downloadable ballistics computers (one Web site lists 28 such programs).[74] When coupled with the fact that defensive measures to prevent a line-of-sight shot from beyond two kilometers are difficult, the constantly improving technology and the established, legal con-sumer community for these technologies make advancements in snip-ing worthy of further examination.

Technological Advance: Ballistics Computers

The development of ballistics computers that are inexpensive,[75] avail-able as software for a wide variety of platforms,[76] and integrated with

---

[72] Examples of manufacturers that provide this sort of information are CheyTac® LLC, Horus Vision LLC, and Trijicon, Inc.

[73] A good example of a Web site that contains enthusiastic articles, tips, a chat area, and even links to Web-based and downloadable ballistics computers is Sniper Country (undated).

[74] See "Ballistics (exterior): Software, Tables and Links" in Sniper Country (undated). Offline, someone hoping to learn more about marksmanship can choose to attend gun shows, shooting demonstrations, sniping adventure camps, and professional courses. For example, it costs £190 (about $335) for a five-hour course at a UK-based gaming venue that offers to teach the following: "All you ever wanted to know about delivering that perfect head shot hit on an enemy target; whether it's a European world leader, a Vietnamese army gen-eral, a terrorist or simply your local traffic warden!" (Combat Games, undated). Professional week-long courses in sniping for under $1,000 are also fairly common. (These courses in particular are limited to military and law enforcement personnel, but not all similar courses are restricted.)

[75] Costs range from no cost at all for something freely downloadable from the Internet (e.g., Web-based programs or the modern Ballistics PC program) to several thousands of dollars for an advanced military product (e.g., CheyTac's $3,000 Advanced Ballistics Computer). See "Ballistics (exterior): Software, Tables and Links" in Sniper Country (undated), Modern Ballistics (undated), and CheyTac (undated).

[76] Although ballistics computers are most commonly available for personal digital assistants (PDAs) and PCs, some computers in novel formats, such as the 5.11 tactical sure-shot calcu-

scopes and other equipment should be considered one of the greatest leaps forward in sniping technology. These computers correct for characteristics of the target, the environment, the shooter, and the weapon and bullet being used. As such, an advanced knowledge of the factors that affect a bullet's trajectory and the ability to perform the necessary calculations to correct for them are no longer prerequisites for successful sniping. Even a basic $100 (or less) ballistics computer can take into account a wide variety of factors (see Table 2.1).[77]

To use a ballistics computer, a sniper need only input the above characteristics. Without access to a computer, a sniper would need a table for each of the factors (e.g., temperature) to allow estimation of

Table 2.1
Capabilities of Sniper Ballistics Computers in Compensating for Target and Environmental Properties

Property	Capability
Target characteristics	Inclination/slant angle (the angle from the shooter to the target) Target speed Target range
Environment	Temperature Barometric pressure Relative humidity Wind speed Wind direction
Shooter behavior	High-end computers can assist in adjusting for activities of the sniper that might affect accuracy
Weapon and projectile	Muzzle velocity Bullet weight Ballistic or C1 coefficient of the bullet Zero range (the range at which the scope was calibrated)

lator watch with software developed by Horus Vision, are now entering the market as well.

[77] These inputs are from the $98 downloadable Horus Vision ATrag1P software for PDAs. See Horus Vision (undated).

its effect on the trajectory, given the range of the shot.[78] The effect is usually related in terms of the windage or elevation minutes of arc (MOAs) adjustment that must be applied to the scope of the rifle to compensate for it (Figure 2.17 shows the windage and elevation adjustments on a common rifle scope). Because there are multiple factors, the sniper must also know how to combine the information read from multiple tables. MOA adjustments are relative to some absolute point, such as sea level when altitude is the factor in question; it is thus also necessary that a sniper be aware of the conditions under which his rifle was zeroed in order to apply the corrections in relative terms.[79]

Figure 2.17
A Common Sniper Scope Features Knobs That Allow a Shooter to Adjust for Elevation and Windage

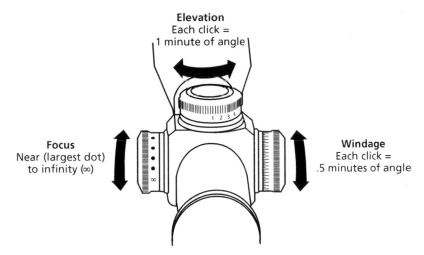

SOURCE: U.S. Department of the Army (1994).

---

[78] Instead of using tables, a sniper can learn certain rules of thumb—e.g., a 20-degree increase in temperature raises the point of impact by one MOA and vice versa—and formulae—e.g., the MOA correction for wind blowing directly from the left or right of the sniper equals (the range of the shot in hundreds of meters the wind speed in miles per hour) divided by a constant C, which is dependent on the range and must be memorized. See U.S. Department of the Army (1994).

[79] See U.S. Department of the Army (1994).

Regardless of whether one thinks about using tables or learning rules and formulae, a computer that removes the need to do either greatly reduces the skills required in successfully hitting a target.

Although basic computers already remove much of the complication in performing a long-range shot, more advanced computers can take many more inputs into account, perform more complex calculations, and include built-in data sets, producing more accurate results easily and quickly (that would be problematic for even the most experienced sniper without the aid of a computer) and pushing even farther the range at which a successful shot is possible. For example, the $3,000 U.S. military software produced by CheyTac Forms Pocket PCs[80] takes into account such factors as the shooter's reaction time, the direction of the shot in relation to true north, the gun's latitude, and the wind speed and direction at three points downrange. It includes a set of live-shot projectile data that was recorded using Doppler radar in a 0–5,000m firing plot and contains information on a bullet's downrange drag and ballistic coefficient in 1-meter increments. Unlike the most basic software, it corrects for phenomena that begin to affect aim at ranges over 1 km, including spin drift and the Coriolis effect.[81]

Although ballistics computers integrate many different types of data and perform complex operations, they are geared toward being easy to use and providing practical outputs. For example, PDA-based computers, whether they are of the basic or advanced variety, usually consist of only a handful of input screens and produce, as output, the scope adjustments that must be made in order to hit the target (see Figures 2.18 and 2.19).[82] In some cases, ballistics computers and scopes have been developed to work together so that no scope

---

[80] A $649 civilian version is available that has some limitations (e.g., it does not include sub-MOA accuracy and its projectile data set has a range of 3,500m instead of 5,000m). Other manufacturers, such as Horus, also produce more advanced computers that will, for example, correct for the Coriolis effect—the sideways drive of a projectile due to the curvature and rotation of the earth—and spin drift—the tendency of a bullet to drive in the direction that its top is spinning.

[81] See "Sniper Country Duty Roster Collective Wisdom" on Sniper Country (undated), CheyTac (2004), and "Coriolis Effect" (2007).

[82] See Horus Vision (undated).

Figure 2.18
The Input Screens on Horus Vision's Basic Ballistics Computer, the ATrag1P,
Are Easy to Understand and Navigate

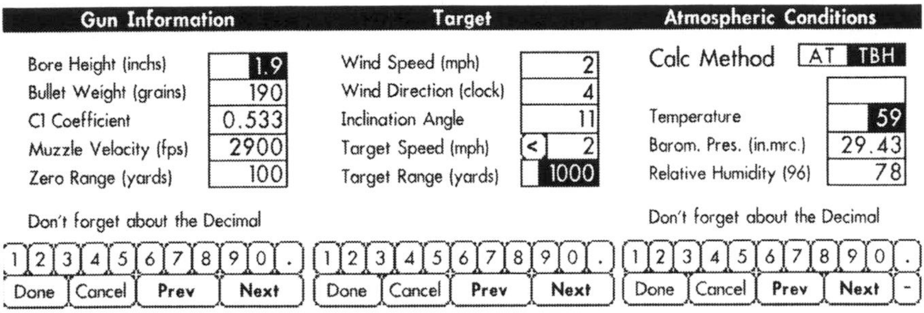

Gun Information		Target		Atmospheric Conditions	
Bore Height (inchs)	1.9	Wind Speed (mph)	2	Calc Method	AT TBH
Bullet Weight (grains)	190	Wind Direction (clock)	4		
CI Coefficient	0.533	Inclination Angle	11	Temperature	59
Muzzle Velocity (fps)	2900	Target Speed (mph)	< 2	Barom. Pres. (in.mrc.)	29.43
Zero Range (yards)	100	Target Range (yards)	1000	Relative Humidity (96)	78

Don't forget about the Decimal   Don't forget about the Decimal

1 2 3 4 5 6 7 8 9 0 .	1 2 3 4 5 6 7 8 9 0 .	1 2 3 4 5 6 7 8 9 0 .
Done Cancel **Prev** **Next**	Done Cancel **Prev** **Next**	Done Cancel **Prev** **Next** -

SOURCE: Horus Vision (2007, p. 11).

Figure 2.19
The Main Screen on the ATrag1P

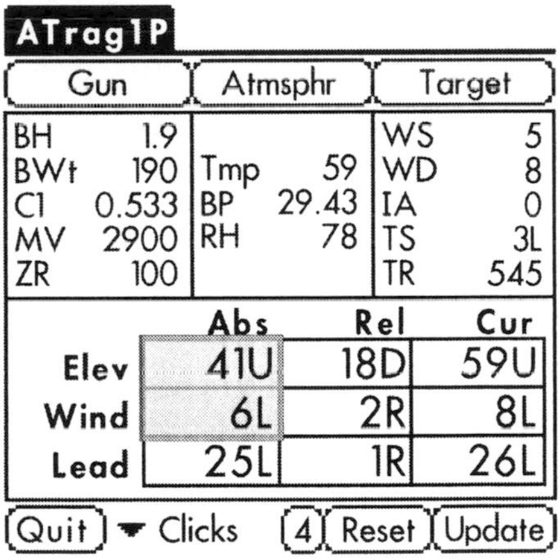

ATrag1P

Gun	Atmsphr	Target

BH	1.9			WS	5
BWt	190	Tmp	59	WD	8
CI	0.533	BP	29.43	IA	0
MV	2900	RH	78	TS	3L
ZR	100			TR	545

	Abs	Rel	Cur
Elev	41U	18D	59U
Wind	6L	2R	8L
Lead	25L	1R	26L

Quit ▼ Clicks   4 Reset Update

SOURCE: Horus Vision (2007, p. 6).
NOTE: This screen clearly indicates the scope
adjustments that must be made to hit the
target, as indicated by the shaded area.

adjustment is needed. One such system is produced by Horus, in which
the scope features a grid-type reticule and the ballistics computer provides the grid coordinates that should be used when aiming at the target
(see Figure 2.20). A target can thus be engaged more quickly than if

Figure 2.20
The Horus Vision Complete Targeting System

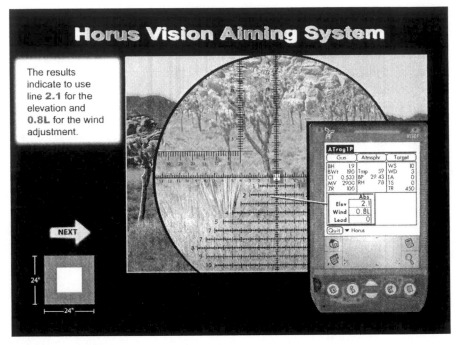

SOURCE: Horus Vision, "The Complete Targeting System: Horus Vision Aiming System,"
slide show demonstration.
NOTE: The system includes a scope with a grid-type reticule and a ballistics computer
that outputs grid coordinates.

the sniper had to adjust the scope and then aim through a traditional
crosshair. Future developments that are being considered by computer
designers and that will further simplify shooting include the ability to
connect sensors in some wired or wireless fashion so that the computer
will automatically read in environment variables.[83] PC-based ballistics
computers often output a wealth of additional information, including
hit probabilities and both simulated tracer arcs and comparative charts
of bullet trajectories. They allow users to practice shots under different
environmental conditions and with various rifles, bullets, scopes, and

---

[83] See CheyTac (undated).

targets. The trajectory data generated in such practice sessions can in some cases be downloaded to PDA versions of the same software.[84]

Technological Advance: Remote Aiming Platforms

Whereas ballistics computers remove the need for an advanced knowledge of ballistics calculations, remote aiming platforms remove the need for basic marksmanship abilities and much of the danger associated with sniping. Although they do not extend the range from weapon to target, they can greatly extend the distance between shooter and prey. One such platform is Precision Remote, Inc.'s Telepresent Rapid Aiming Platform (TRAP)—see Figure 2.21. The components

Figure 2.21
The Components of the TRAP System

SOURCE: Courtesy of Precision Remotes.

---

[84] See Modern Ballistics (undated).

of this system that fundamentally change the practice of sniping are as follows:

- a tripod-based recoil- and accuracy-stabilized platform for weapons up to 0.50 caliber that can be mounted on a moving vehicle or in a fixed location
- through-sight and surveillance cameras linked to video monitors (with the through-sight system providing a resolution of 0.2 MOA)
- remote operation via cable at a range of up to 100m or through a wireless connection at a range of up to 1 km
- the ability to rotate on horizontal and vertical axes and pan sufficiently quickly to track a crossing target at 100m that is moving at 48 km per hour
- the ability to provide automatic correction for range and wind conditions.

The existence of a system with these types of components means that a sniper could shoot a target in relative safety with little risk of being caught and without ever having practiced such fundamentals of marksmanship as finding a steady position and breathing, pulling the trigger, and following through in such a way as to minimize the disruption to the rifle's position. It is not only an accurate system, but one that is easy to learn: With estimated training times of one hour for basic competency and 10 hours for expertise, it is certainly much faster than traditional training.

This type of technology is already in production and has been delivered to such groups as the Israeli Defense Force, all branches of the U.S. Department of Defense, and the U.S. Department of Energy.[85] Although it has been fielded by government entities, it does not seem to be available for purchase by civilians. Thus, though the ease of use and remote operation offered by the TRAP could make it attractive to many would-be attackers, it may be harder to obtain than other sniping equipment.

---

[85] See Precision Remotes (undated) and Daniel (2004).

Though neither as advanced nor as versatile as TRAP, shooting benches, costing under $100 and freely available, offer some advantages. Most important, they allow the sniper to clamp the gun into a position and so eliminate some sources of misses. Of course, they also deny the sniper an ability to react to the target. Still, a less well-trained terrorist shooting at a fixed location could find such a bench useful.

Technological Advance: Enhanced Scopes and Reticules

When compared to an iron sight or a stock scope such as the M3A used by the U.S. Army in the 1990s,[86] breakthroughs in sighting technology have made it easier to find a target quickly and aim at it accurately. Some of the breakthroughs have come in the form of superior integration with ballistics computers (discussed above), or with night-vision and rangefinder devices, discussed in sections that follow. Other breakthroughs have come about due to a better understanding of human vision and what is needed to perform the complete act of first finding the target and then aiming at it. Unmagnified red-dot sights, for example, have been fairly common for some time and have caused a lighted reticule to appear as if it were projected onto the target. When using such a sight, the shooter does not need to close one eye and can take advantage of being able to see the entire scene with regular depth perception in order to find the target more easily.[87]

An evolution of red-dot sighting is the Bindon Aiming Concept (BAC), in which a red-dot–style reticule is used and both eyes are left open to pick out the target but the scope is magnified (see Figure 2.22). The BAC relies on research that showed that, when a rifle with a magnified scope is moved to find a target, the brain focuses on the image it receives from the unaided eye instead of the blurry image with the floating reticule that it receives from the eye that is looking through the scope. Then, when the target is found and the rifle stops moving,

---

[86] The only aid provided in the m3A is a mil-dot reticule (the mil-dot system consists of dots one milliradian apart on the crosshairs that allow the shooter to get an estimate of target range if the target is of known size). See U.S. Department of the Army (1994) and "Mil-Dots and Minutes of Angle" on Sniper Country (undated).

[87] See "Red Dot Sight" (undated).

Figure 2.22
The BAC Allows Both Eyes to Stay Open While Using a
Magnified Scope

SOURCE: Courtesy of Trijicon.

the brain switches to the magnified view. The BAC thus combines the speed of red-dot sighting with the precision of a telescopic scope.[88] It is an example of how the line between combat and hunting equipment is blurring: Scopes incorporating BAC were developed for the military and only recently became available to hunters. These types of innovations that are increasingly available to nonmilitary consumers not only improve a hunter's chances of felling the quarry, but also could prove useful to tactical shooters engaging human targets.

Technological Advance: Long-Range Night Vision for Snipers

Light intensification and thermal imaging technologies are increasingly finding their way into a number of different tools for snipers. Cooled thermal and Gen-III/Gen-IV[89] light intensification sights are available, for example, either as clip-on devices for use with an installed daytime scope or as stand-alone scopes that can be used only at night. The advantage of a clip-on product is that a sniper can continue to use the scope and reticule with which he or she is familiar and can easily

---

[88] Trijicon (undated).

[89] Please see the separate section on developments in light intensification and thermal devices for an explanation of these technologies.

switch between daytime and nighttime shooting. A stand-alone sight is usually more powerful and has more features. With newer stand-alone sights, a shooter can see out to ranges of 1,500m, change screen colors and reticules entirely through software, and transmit output to video. In addition to these sights, a number of devices that are handheld or worn as goggles are being fielded that combine night vision with other features such as laser rangefinders, GPS, and digital compasses.[90]

Although the use of Gen-III and Gen-IV intensification technologies is supposed to be limited to certain militaries, anyone can buy hunting scopes using Gen-IV tubes.[91] Easily acquired night-vision devices that allow one to see as clearly in nighttime as in the daytime and far enough to perform a long-range shot (out to 1.5 km) potentially give a sniper the ability not only to shoot from beyond a customary perimeter of security but also to do so in the cover of darkness.

Technological Advance: Rangefinders

Because the effects of gravity, weather conditions, bullet drag, and other phenomena such as spin drift vary with range, a basic skill traditionally needed by snipers is the ability to calculate the range to the target using some form of in-reticule system (e.g., the mil-dot system). Calculating successfully not only requires knowledge of and practice with a measurement system, but also an ability to judge the target size, which can become tricky with angle shooting.[92] Performing this calcu-

---

[90] "Shoot-Out at Blackwater" (2005).

[91] See American Technologies Network Corporation (undated).

[92] Most marksmen will practice shooting on the same horizontal plane as their target and often directly in front of it (many firing ranges are set up for this head-on style shooting). Shooting in this way and knowing the size of the target (e.g., a standing man measures about 1m from crotch to head), a marksman can see how many units of measurement in the reticule (e.g., mil-dots) are covered by specific portions of the target and estimate its range with a few learned formulae. Examples of these formulae when "mil'ling" (using the mil-dot system for range estimation) humans include 1,000/mils = range in m (for a crotch-to-head estimate of the mil-dots covered), 500/mils = range in m (for a shoulder-to-shoulder estimate), and 250/mils = range in m (for a side-to-side-of-the-head estimate). Additionally, the sniper needs to know not to use the side-to-side-of-the-head estimation for targets more than 400m away, not to use shoulder-to-shoulder past 600m, and that all mil-dot use becomes difficult past 1 km (see Haugen, 2001). Angle shooting, the act of shooting below

lation takes time, a valuable commodity if the target is available only briefly. Laser rangefinders, which work by measuring the time it takes for a laser signal to bounce back from a target, do away with the need for a sniper to calculate distances and provide near-instantaneous range information. The speed with which the information is provided can be especially important in situations in which a target is moving and its distance from the shooter changes rapidly.

Recent advances in rangefinders include increased range and accuracy capabilities, the miniaturization of components, and integration with scopes, compasses, and GPS. Many popular laser rangefinders available to civilians for hunting have a maximum range of about 1 km, although accuracy degrades with range.[93] Rangefinders can be found in pocket-sized devices, binoculars, and even integrated with scopes (as described above).[94] The instruments with which they are combined are also changing. The utility provided by combining a rangefinder with a compass is that the vector instead of the scalar range (the vector includes not only the distance but also the direction) to the target can be provided to the sniper. When this vector is further combined with GPS and the coordinates of the shooter are known, the exact coordinates of the target can be calculated as well.[95]

Technological Advance: Environmental Sensors

A bullet's trajectory is affected by barometric pressure, humidity, temperature, and wind speed. Sensors that measure all of these environ-

---

or above the target, introduces further complications into performing the shot successfully. One of these complications is that it becomes harder to calculate range using an estimate of a target's size and the known formulae described above. Additional math would be necessary. Another complication is that the ballistics trajectory is affected by the angle of the shot. The change in the trajectory could be corrected for using a ballistics computer like those described above.

[93] For example, the Leica LRF 1200 Rangemaster has a ± 0.9m accuracy at 400m, ± 1.8m accuracy at 800m, and ± 0.5 percent of the range beyond 800m.

[94] See Adorama (undated) and Sundra (1999).

[95] Rice (2004).

mental characteristics are currently available in inexpensive all-in-one handheld units such as the $300 Kestrel® 4000 (see Figure 2.23).[96]

Such units make it unnecessary for a sniper to know how to "read" the environment. For example, traditionally trained snipers could be

Figure 2.23
The Kestrel 4000 Is Small, Cheap, and Useful to Snipers

SOURCE: Courtesy of Nielsen Kellerman.

---

[96] Nielsen Kellerman (undated). Kestrel® is a registered trademark of Nielsen Kellerman.

expected to perform such tasks as watching a flag and dividing the angle between the pole and flag by four to estimate the wind speed in miles per hour.[97]

Because wind speed and direction change downrange, however, all-in-one units that provide the sniper only with weather data solely at his or her location may not provide sufficient accuracy for a long-range shot. The current work-around used by the sniping community is to take readings at several points downrange using remote anemometers.[98]

These remote sensors may soon be replaced by laser anemometers, the feasibility of which is currently being studied by the Canadian and other militaries. Such a laser device would be operated from the shooter's location but be capable of measuring the crosswind velocity at several points downrange.[99]

Antitank Guided Weapons

This class of infantry weapons began to appear in the early 1970s. The Yom Kippur War of 1973 among Israel, Egypt, and Syria marked the first large-scale use of these weapons. Initial interpretations of that war included spectacular predictions that the use of wire-guided antitank missiles had rendered the tank obsolete. Although those statements proved to be exaggerated, ATGW remain formidable systems that can also be used against buildings, unarmored vehicles, and bunkers.

Compared with RPG-type weapons (which are covered in the next section), ATGW are larger, usually requiring a crew of two to four personnel to carry the launcher and ammunition (see Figure 2.24). Most ATGW are wire guided; the operator keeps his or her sight fixed on the target after missile launch, and correction commands are transmitted to the missile along the guidance wire. The most modern ATGW are fire-and-forget systems with no wire trailing behind the missile; once

---

[97] U.S. Department of the Army (1994).

[98] CheyTac (2001–2006).

[99] Nappert, Champagne, and Taillon (2002).

**Figure 2.24
Russian-Built AT-4 Spigot, an
Antitank Guided Missile and
Launcher**

SOURCE: Courtesy of U.S. Department
of Defense.

the operator has locked the missile onto the target, the weapon is fired and seekers aboard the missile maneuver it to the target. In all cases, the system consists of a launcher and control unit and the ammunition. Typically the ammunition is "clipped into" the launcher and control unit. Once a missile is fired, the gunner's assistants ready another missile by clipping it into the launcher and making the required electrical connections between the missile and the launcher and control unit.

Ranges of ATGW vary from roughly 1,000 meters (example: the U.S.-made *Dragon* of the 1970s) to more than 4,000 meters (the U.S. *Javelin* or the European *HOT*). The warheads are usually an antiarmor shaped charge that is optimized for armor penetration. Under optimal conditions, modern, large ATGW can penetrate more than 600mm (roughly 24 inches) of solid, rolled homogeneous armor plate. The large warheads also have considerable blast effect, thus endangering nearby personnel or unarmored vehicles that are near the impact point. The warheads of some modern ATGW are designed to detonate above the target, with the blast effect directed downward against the vulnerable top of the armored vehicle target.[100]

From a terrorist group's viewpoint, ATGW could provide a powerful, highly accurate weapon with practical ranges of one to two miles or more. The antiarmor warheads of ATGW, being designed to defeat heavily protected main battle tanks, can easily destroy any civilian-type vehicle in the world, including armored trucks and limousines. Although primarily designed for use against armored vehicles, ATGW have frequently been used in combat against buildings and bunkers. For example, the Israelis have often used U.S.-designed Hellfire helicopter-launched ATGW to destroy buildings of opposition groups such as the Palestinian Liberation Organization and Hamas. The Hellfire is so accurate that specific windows on a building can often be targeted. The terrorist application of such systems is obvious.

Man-Portable Antiarmor Weapons

The most common examples of man-portable antiarmor systems are the RPGs of the type so often used by the insurgents in Iraq. Easily carried by one person (although others may need to carry additional ammunition), these weapons originated during World War II to give the infantry a viable, short-range antitank capability. Today this class of weapon has an effective range of 300 to 500 meters, depending on whether the target is moving or stationary. The most common type of warhead available for this class of weapon is the antiarmor shaped charge (in U.S. terminology, this warhead is often called HEAT, for high-explosive antitank). Although shaped charge warheads are optimized for armor penetration, they also have a blast effect when they explode, posing a threat to exposed personnel near the blast. In addition to the normally shaped charge warheads, it is typical for these weapons to have other warhead options such as high explosives. Although the HE warhead (as opposed to shaped charge) is relatively ineffective against armored vehicles, it can be devastating when used against

---

[100]"Anti-Armor Missiles" (2005).

unarmored vehicles such as cars and trucks. Additionally, the HE versions produce greater blast effect, making them better weapons against exposed personnel.

Terrorist organizations have been using this class of weapon for many years. Indeed, the Soviet-designed RPG-7 (see Figure 2.25) has become ubiquitous among terrorist, insurgent, and militia organizations in much of Africa and the Middle East. Cheap and easy to use, man-portable antiarmor weapons are already a major element in the arsenal of terrorist groups.

The next generation of man-portable antiarmor weapons will be even more effective. Some of the ongoing improvements to this type of weapon include

- enhanced computer-aided sights
- more warhead options (including smoke, fragmentation, and airburst)
- the ability to soft-launch from inside a building (earlier versions had large amounts of back blast, making firing them from inside buildings very dangerous to the gunner)
- longer ranges, out to roughly 1,000 meters.

The very latest weapons of this type include guided projectiles, although this capability is still rare today. Guided projectiles will be launched by the gunner in the direction of the target. After launch, the warhead guidance system will track the target and correct the

Figure 2.25
The RPG-7 Antitank Grenade Launcher

SOURCE: Courtesy of U.S. Army Signal Center.

projectile onto the target. These projectiles are actually powered missiles, as opposed to the unguided, unpowered warheads of the current generation of man-portable systems. Examples of this new trend include the British-Swedish MBT LAW (main battle tank and light armored weapon).[101]

Limpet Mines

These weapons are devices that are attached to the underwater portion of a ship's hull, to a pier, or to other platforms such as oil rigs. In use since before World War II, limpet mines range in size from less than 10 kg to several hundred kg. Today, many nations produce limpet mines of various sizes and capabilities. These devices are simple in concept and design. Relatively little training would be required to employ most modern limpet mines. The time required to prepare underwater swimmers to use their scuba equipment would probably be much more extensive than the training needed to employ the mine successfully.

During World War II, naval commandos of several nations—notably Italy and the UK—successfully employed limpet mines such as the one shown in Figure 2.26.[102] The weapons were usually delivered and emplaced by divers who were taken close to their targets by submarine. Once they had left the transport submarine, the divers would either swim to their targets or be carried there by self-propelled underwater devices that generally resembled torpedoes. Once near their targets, the divers would emplace the limpet mines on the hull of unsuspecting warships or merchantmen, set timers, and then attempt to return to their transport submarines. On several occasions, limpet mines sank large ships in harbors. For example, on December 18, 1941, during World War II, two UK battleships were sunk in Alexandria harbor in Egypt by Italian divers.[103]

---

[101] "MBT LAW Light Anti-Tank Weapon" (undated).

[102] Trueman (undated).

[103] Giammario (undated).

**Figure 2.26
Magnetic Limpet Mine**

SOURCE: Courtesy of Royal Marines Museum.

The mode of attack used during World War II is very similar to the one in which limpet mines would be employed today. Underwater swimmers would attempt to reach their targets undetected. Modern limpet mines are usually attached to their targets magnetically and include antitamper devices that will cause them to explode if an attempt is made to remove them.

For a swimmer to effectively deliver a limpet mine, the device has to be relatively small. Large, bulky devices would cause considerable water resistance that would slow down an underwater swimmer. Indeed, in most cases, the dimensions of a limpet mine would be more of a challenge to an underwater swimmer than the weight of the mine would be. This means that the practical size limit of most swimmer-carried limpets would be in the 10 to 20 kg weight class, thus restricting the explosive effect of the device. While small vessels (approximately the size of a tug boat or small vehicle ferry) might be badly damaged or sunk by a 10 to 20 kg mine, several weapons would be needed to inflict major damage to a large cargo ship. An example of such an attack on a small vessel was the 1985 French attack on the *Rainbow Warrior* in New Zealand.[104]

If an underwater transport assist is used, however, much larger devices can be used. Similar in concept to the World War II–era underwater transport platforms, a self-propelled underwater carrying device would allow for limpets of many hundreds of kilograms to be moved to a target. One such "chariot" or swimmer delivery vehicle is the Italian CE2F/X100. It is reported to have the ability to carry a mine with 230 kilograms of explosives and two divers up to 50 nautical miles and to be in service in Argentina, India, and Pakistan.[105] Devices with warheads of several hundred kilograms of explosives are available today for naval commandos of several nations. Such devices would allow terrorists to inflict major damage to, or sink, even large vessels of 10,000 tons or more.[106]

Advanced Land Mines

From the end of World War I to the 1970s, the vast majority of land mines were simple, "dumb" systems that were emplaced just below the surface of the ground. Two general types of mines were used—antipersonnel and antitank. In recent years, the trend has been toward "smart" mines that can discriminate between targets; have antitamper devices; and can, in some cases, attack targets from a stand-off distance of many meters. Although some mines are still buried underground for concealment, many of the new smart mines are placed on the surface and camouflaged, since their sensors and firing devices cannot oper-

---

[104]Hunter (1995).

[105]See *Jane's Underwater Warfare Systems* (2005).

[106]See "Mini Submarines and Special Forces Pose Maximum Threat" (1998).

ate while buried. Examples of today's advanced mines include systems with acoustic sensors that can be hidden several meters off the side of a road. When the mine "hears" the right type of vehicle sounds, the mine launches a "pop-up" device that attacks the passing vehicle.

Similar technology is used in antihelicopter mines that several countries are developing. The mines are intended to attack relatively slow, low-flying air vehicles, primarily helicopters, by listening for them, then initiating an attack when the aircraft is within range. Although the development of antihelicopter mines (which was under way in several countries in the 1990s) has slowed in recent years, several versions are still reported to be in development around the world.[107]

Night Vision

There are two main technologies that can be used to see in the dark that are undergoing constant improvement: light intensification, which amplifies the existing visible light; and thermal imaging, which creates an image from the heat radiated by an object. Image fusion attempts to combine both technologies for maximum benefit.

Technological Advance: Four Generations of Light Intensification

Improvements in light intensification have mainly focused on reducing the fuzziness that occurs when the gain is increased. In fact, intensification devices are categorized into generations based on these fuzziness-eliminating improvements. The 1960s Gen-I devices use electrostatic tubes to turn photons into electrons, multiply them, and project them onto a screen. The 1970s Gen-II devices use a microchannel plate to channel and further multiply the electrons, resulting in a sharper, brighter image. The 1980s Gen-III devices include a film deposited on the microchannel plate to protect the plate.[108] Finally, modern Gen-IV "gated filmless" technology does not use film and instead preserves the

---

[107]"Anti-Helicopter Mine Displayed at HEMUS-96" (1996), "Anti-Helicopter Mines" (2005). See also *Jane's International Defence Review* (1998).

[108] Biass (2004a).

life of the plate and produces a clearer picture by switching the electron source on and off at a very high rate (it was found that the film in Gen-III systems deteriorated with use). Gen-I technology is still found in the most inexpensive consumer products, but later generations of the technology are also available within the United States,[109] even though their export from the United States is controlled by Category XII of the International Traffic in Arms Regulations.[110]

Technological Advance: Seeing Heat

When it comes to developments in thermal imaging, much of the focus has been on improving the sensitivity and element density of infrared sensors to improve the contrast and resolution of the images they create. The desire to make more portable sensors, on the one hand, but more sensitive ones, on the other, has led to the creation of two types of thermal imaging devices: cooled and uncooled.

Cryogenically cooled sensors are more sensitive and can detect differences in temperature as small as 0.2 degrees Fahrenheit but are less portable and more expensive. A major benefit of all thermal imagers over light intensifiers is their ability to provide information about the environment that would not even be visible in daylight (e.g., it can reveal concealed living targets, whether or not a vehicle's motor is or was recently running). Because of this benefit, there has been interest in fusing light intensification and thermal imaging technologies. A prototype of a scope that accomplishes this fusion was developed in FY 2002.[111]

---

[109] Although the sale of Gen-IV night vision is supposedly restricted to U.S. forces, it is possible to order a fourth-generation scope that uses "current military designs" and is "the ultimate choice for Law Enforcement Professionals or the Hunter that insists on his equipment being the best." See American Technologies Network Corporation (undated). Purchasers are warned that "export of the commodities described herein is strictly prohibited without a valid export license."

[110] "Mortars" (1998).

[111] For several examples, see Aurora Tactical (undated), Laser-King Companies (undated), Biass (June 2004b); U.S. Department of Defense (2003).

CHAPTER THREE

What Advanced Conventional Weapons Are Potentially Most Useful and Attractive to Terrorists?

In assessing the attractiveness of a particular advanced conventional weapon system to terrorist adversaries, the central consideration must be the potential benefit that the terrorist organization might gain by acquiring and using the weapon. Having examined a variety of next-generation weapons, it is clear that there are many potentially dangerous conventional weapons being introduced into the inventories of modern military forces around the world. However, most of the weapons improve upon the lethality, usability, or reliability of existing weapons on the margin. As such, these new weapons are excellent candidates for replacing obsolete weapons but, from a homeland security perspective, which is focused on concerns about the potential impact of small numbers of these weapons used by terrorists, they do not dramatically alter the potential impact of terrorist operations. Only a few of the weapons under development stand out as what we have termed *game-changing weapons*—systems that fundamentally alter the relationship between the attacker and the defender.[1] Such major shifts are of the most concern with respect to homeland security in general and in the design of protective measures specifically.

---

[1] Dramatic enhancements that change the probability of arrival, probability of hit, or probability of kill all have the potential to be game-changing weapons. In the context of conventional weapons, the greatest changes seem to be associated with altering the probability of arrival or the probability of a hit.

Game-Changing Weapons

A game-changing weapon falling into the hands of a would-be attacker would be one that would force the defender to dramatically alter its behavior to counter these new weapons. As a result, such weapons represent cases that would be of greater interest in terms of focusing future efforts to control the availability and usability of these weapons by terrorist groups. The effects of game-changing weapon systems can be explored by thinking about the potential target set that a terrorist organization might be able to threaten to use in a homeland security context.

This potential target set can conveniently be thought of as being composed of people and things located in one of three different kinds of sites (see Figure 3.1).[2]

- The first consists of sites completely accessible to the general public in the course of daily activities. These locations might have some restricted elements (i.e., closed areas), but the vast majority of elements are directly accessible to the public. The infrastructure and people immediately surrounding that point are easily accessible to would-be attackers and consequently are vulnerable to the full array of close-in weapons such as bombs or small arms.
- The second category consists of targets that somewhat limit access. This access limitation would be consistent with many industrial areas and with some limited-access government facilities with perimeter security. These installations are more difficult to gain access to, and perimeter security would need to be directly breached or bypassed in some manner to strike a target.
- The final category is that of restricted-access sites. These sites have effectively no routine access by the public; security is much tighter; and, frequently, quick-reaction security forces are available to deal with penetrations. Any attacker would have to deal

---

[2] A similar approach was used originally in Baker et al. (2004). The study focused on the ability to gather intelligence on potential targets as a function of the ability to gain direct access to the target area. Huge gains were seen from approaching a target for intelligence, and similar gains are seen when considering attack opportunities.

Figure 3.1
Potential Terrorist Targets by Degree of Public Accessibility

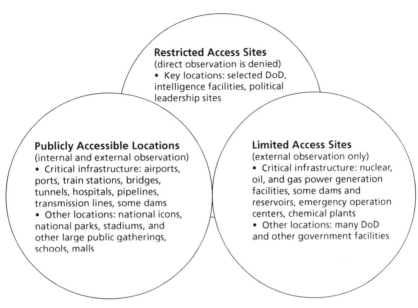

Restricted Access Sites
(direct observation is denied)
• Key locations: selected DoD, intelligence facilities, political leadership sites

Publicly Accessible Locations
(internal and external observation)
• Critical infrastructure: airports, ports, train stations, bridges, tunnels, hospitals, pipelines, transmission lines, some dams
• Other locations: national icons, national parks, stadiums, and other large public gatherings, schools, malls

Limited Access Sites
(external observation only)
• Critical infrastructure: nuclear, oil, and gas power generation facilities, some dams and reservoirs, emergency operation centers, chemical plants
• Other locations: many DoD and other government facilities

RAND *MG510-3.1*

with not only the perimeter and immediate security forces, but also with quick-reaction forces that could severely limit the ability to move to the intended target area.

Of the myriad of factors that might matter to the defender, a subset of these that most strongly alter the probability of success against a single target are clearly very important to a potential attacker, and consequently, they represent an efficient means of identifying what would be high-leverage weapons from their perspective, such as the probability of the weapon being moved successfully into strike position, the probability of the weapon arriving on target, the probability of a hit upon arrival, and the probability of a kill after a hit—and, to a much lesser extent, the probability of escaping from the target area to facilitate reattack operations. Of all the factors, the most significant would seem to be the ability to strike at targets that heretofore had been offered some degree of protection either by security, construction, or other factor. In general, most of the weapons described previously do not have the greatest benefits when striking at publicly accessible, "open" targets—although they may provide some advantage, the effectiveness of low-technology methods of delivering explosive devices would limit the marginal increase in the threat posed by a terrorist group. Where the weapons have a more substantial impact is in the striking of limited- and restricted-access sites that primarily depend on perimeter security for protection against attack.

Terrorist Scenarios Involving Advanced Conventional Weapons

To better understand which of the advanced conventional weapons represented game-changing capabilities for terrorist adversaries and, therefore, were potentially attractive weapons from their point of view, the research team explored a range of scenarios. From a broader examination of the weapons discussed in the previous chapter, this analysis identified the subset of particular concern. This examination focused on the use of advanced mortar systems, sniper weapons, advanced small arms, ATGW, and limpet mines.

Mortar Attacks

Mortars have recently gained widespread media attention—ever since insurgents in Iraq used them to try to disrupt the January 30, 2005, elections[3] and to attack targets of all sorts, ranging from protected military structures such as U.S. bases to unprotected civilian buildings and infrastructure such as fire stations, colleges, and telephone exchanges. An illustration of the frequency of such attacks can be seen in the nickname given to Camp Anaconda, the largest U.S. support base in Iraq: Mortaritaville.[4] The widespread use of mortars in Iraq demonstrates

---

[3] The RAND-MIPT Terrorism Incident Database lists no fewer than 32 mortar attacks against Iraqi polling stations and voting centers in January 2005.

[4] "Balad Airbase" (undated).

that organizations targeting U.S. interests see value in adopting this class of weapon and that they have vigorously done so.

The situation in Iraq is not unique, however. Both the past and contemporary history of terrorist activity worldwide demonstrates that a range of violent groups has adopted mortars as a weapon to provide indirect fire capabilities and therefore has a significant body of experience with these weapons. Data available in the RAND Terrorism Chronology and in the RAND-MIPT Terrorism Incident Database[5] show the following:

- **Broad use of mortars:** There have been recorded instances of mortars in terrorist attacks by at least 30 separate terrorist organizations, ranging from large and sophisticated groups such as the Liberation Tigers of Tamil Eelam (LTTE) and Hizballah to groups that are essentially unknown outside their theaters of operation, with much lower capability levels. Such broad use of the weapon—across the full spectrum of terrorist groups, from ethnonationalist to jihadist groups—demonstrates the utility and attractiveness of this class of weapons in many operational contexts.

- **Extensive recent use:** Terrorists have used mortars all around the world, but available data show most extensive use in the ongoing conflicts in the Middle East, both by the Iraqi insurgency and by Palestinian groups carrying out operations against Israel. Out of the 542 mortar attacks occurring since 1998 (on which data are available in the RAND-MIPT Terrorism Incident Database), 503, or 92 percent, were staged in the Middle East and Persian Gulf region (see Figure 3.2). Beyond the occurrence of the events themselves, the extensive recent use of these weapons is troubling from a proliferation perspective. These conflicts have helped to build a group of individuals who are familiar with these weapons and have used them in combat situations.

---

[5] The RAND Terrorism Chronology records international terrorist incidents that occurred between 1968 and 1997, while the RAND-MIPT Terrorism Incident Database records domestic and international terrorist incidents that occurred from 1998 to the present.

Figure 3.2
Mortar Attacks Since 1998, by Region

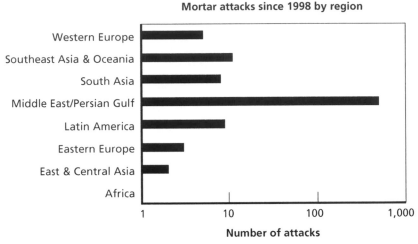

SOURCE: National Memorial Institute for the Prevention of Terrorism and RAND
Corporation (ongoing).
RAND *MG510-3.2*

- **Potential for anonymity and escape:** Although mortar attacks
  have been attributed to a significant number of individual terrorist groups, the perpetrators of the vast majority of recorded
  attacks are unknown. Out of the 542 mortar attacks occurring
  since 1998, the perpetrating organization is unknown for 337 or
  62 percent of them (see Figure 3.3). This anonymity derives from
  the ability to deliver indirect fire from a distance away from a
  target, potentially coupled with the ability to trigger these weapons remotely. The relatively small size and weight of mortars make
  them easy to transport and conceal. Even a 120mm mortar can
  easily fit into the rear of a pickup truck or SUV-type vehicle,
  along with several rounds of ammunition. The same characteristics can allow a terrorist to escape capture, perhaps with some or
  all of the mortar equipment. This ability to fire and yet escape to
  fight another day is likely one reason for the popularity of these
  weapons, particularly in high operational tempo situations during
  ongoing conflicts. As technology advances make mortar systems
  more lightweight, easier to use, and capable of longer ranges, the

Figure 3.3
Mortar Attacks Staged Between January 1998 and June 2005

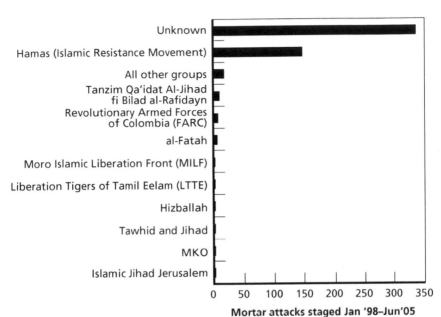

SOURCE: National Memorial Institute for the Prevention of Terrorism and RAND Corporation (ongoing).
RAND *MG510-3.3*

potential for anonymity will increase. These characteristics are not unique to terrorist attacks using mortars. In fact, the proportion of anonymous attacks for terrorist bombings is similar. Remotely controlled bombs also offer an opportunity for escape. Perhaps not surprisingly, bombs are also commonly used by terrorist groups.

- **Restricted effectiveness of terrorist mortar use:** In spite of the extensive use of mortars by terrorist groups, attacks using these weapons do not frequently produce large numbers of casualties. In the 542 attacks discussed previously, only 267 people were killed, or approximately one person for every two attacks. Although the indirect fire capability provided by mortars is an effective terror weapon whether or not attacks reproducibly strike their targets or kill their intended victims, limits on how well the weapons have been used have placed a bound on the outcomes that terrorists have been able to achieve by using them. These shortcomings have originated both from the fact that most terrorist groups have used improvised rather than commercially produced mortars (where design and manufacturing shortcomings reduced their effectiveness) and from operational limitations. If terrorists expect a clean escape from the scene of the attack, they cannot fire multiple volleys of mortar shells; for this reason, they lack the opportunity to correct for any shortcomings in the aiming of their initial shots.

The experience of the Provisional Irish Republican Army (PIRA), long considered the experts on mortar use among terrorist groups, is instructive in this regard. During its operations between 1969 and the mid-1990s, PIRA staged large numbers of mortar attacks. By many measures, PIRA's use of mortars for much of this period appears to have been only marginally successful. In most of PIRA's attacks, the casualties were relatively low. "Between 1973 and early 1978, they attempted a total of 71 attacks using such weapons without killing a single member of the security forces."[6] Incident data reported by the PIRA-associated publication *Iris* for the years 1981 and 1992 include approximately 92 mortar attacks launched during the time period (not counting any that were discovered and defused by security forces). In spite of the likely reporting biases in the source, the magazine indicated that only 9 percent of those attacks resulted in fatalities and that, for 35 percent of them, there were no reported casualties at all. PIRA also staged high-profile mortar attacks that "just missed" because of errors in laying or firing their mortars. For example, PIRA's attack on 10 Downing Street[7] would have achieved a direct hit on the prime minister's council chambers during a cabinet meeting had the mortar been laid and angled slightly differently. Sometimes mortar shells were significantly off course, missing intended targets entirely (and produc-

---

[6] Urban (1992), p. 206.

[7] Bell (1998), p. 192.

ing collateral damage embarrassing to the group) or injuring group members.[8]

The capabilities of new mortar systems that address the shortcomings in gun-laying and use that limited the effectiveness of previous generations of mortars and improvised systems could increase the attractiveness of these advanced mortars to terrorist groups. Two types of terrorist attacks taking advantage of some of the advances in mortar technology can be envisioned. In one scenario, terrorists could make use of longer-range mortar bombs and perhaps the ease of use offered by a fire control system and a GLS to attack a large secured area. In another scenario, they could use one of the man-in-the-loop terminal guidance warheads for the precision needed to carry out an assassination. Ultimately, a fully capable, GPS-only system would avoid the need for man-in-the-loop guidance entirely, lowering the complexity of the attack plan dramatically.

Attack on the Rose Bowl: A Scenario

Every year on the first of the year, up to 90,000 enthusiastic fans fill the Rose Bowl in Pasadena, California, to usher in the New Year by watching two of the best college football teams take each other on. An additional 425 million spectators nationally and 164 million internationally tune into the game on ABC and ESPN.[9]

Fast forward to January 1, 2008. As fans pass through security, bags are inspected and a handful of people are selected for pat-down searches. Meanwhile, five miles to the west of the stadium, three men drive their white van into Lower Scholl Canyon Park in the hills surrounding Pasadena and park behind a ridge in a secluded area (see Figure 3.4). Alternatively, they drive north of the stadium and park 10 km away in the San Gabriel Mountains. Nervous even though they used mortars regularly when fighting as insurgents in Iraq, they are a bit clumsy as they pull out three components of a U.S. Army M120 mortar system, including the M298 cannon assembly (110 lbs), the M190 bipod assembly (70 lbs), and the M9 baseplate (136 lbs). It takes them five minutes to assemble the pieces. They

---

8   Bell (1998), pp. 191–192; O'Callaghan (1999), pp. 84–85.

9   See Pasadena Tournament of Roses (undated).

Figure 3.4
Potential Baseplate Site

SOURCE: Courtesy of Google Earth™.
NOTE: These are two potential baseplate sites—hills 5 km west of the Rose
Bowl and mountains 10 km north.
RAND *MG510-3.4*

then return to the van and pull out four XM984 extended-range cartridge munitions (ERCMs) capable of an 11 km range. Using an azimuth gun-laying system[10] *and knowing the GPS coordinates of both their location and a section of the stadium bleachers from having previously staked them out, they take another five minutes to set the angle and orientation of the tube. They set the fuzes on the mortars, which determine the delay before the submunitions are released, and they drop the first mortar round into the tube. Within one minute, all four are launched.*

---

[10] Presently, of the XM32 lightweight handheld mortar ballistic computer is being developed, which is not capable of weapon pointing, but handheld AGLS/MFCS devices exist for many other mortar launchers, and there is a push to develop a lightweight one for the dismounted 120mm U.S. system. (See Gourley, 2004.)

As the marching band files off the field and is replaced by the players, the thunderous applause of the audience is replaced by the explosion of the 54 M80 grenades dropped by each of the four mortars, each volley of 54 covering an area more than twice that of an HE warhead. Dozens of fans are dead before anyone knows why. Nobody thinks to duck behind his or her seat for protection. After a stunned pause, dozens more are killed, trampled in the ensuing rush to the exits. Millions in the United States and around the world are glued to the live broadcast as they watch the carnage unfold before their eyes. Many cannot believe that a tradition in the United States' most traditional sport has turned into a scene of horror.

When the abandoned mortar tube is found in the hills many hours later by FBI agents, there is no one in sight. Days later, as announcements continue to be made about the canceling of outdoor and stadium events over fears of more attacks, an al Qaeda group announces on its Web site that it takes credit for the attack. Without ever needing to smuggle any sort of weapon into a secure location, a handful of people succeed in attacking the United States—symbolically and in actuality—causing both immediate physical harm and psychological effects that ripple throughout the nation's economy and collective psyche for years to come.

The use of an advanced system by terrorists, many of whom are currently gaining experience with traditional mortars in Iraq and the Occupied Territories in Israel, could allow terrorists to hit a protected target from many kilometers outside any reasonably defensive perimeter and from behind the cover of topographical features such as mountains or forests. Because of the speed with which new mortars can be fired and their increased range, there is a high probability that terrorists could unleash a barrage of projectiles on their targets and escape undetected.

The size and general simplicity of mortars should also make them surprisingly easy to smuggle into the United States, especially if they were broken down into their component parts. Mortars have four main components, exclusive of their ammunition: the barrel that, to an untrained eye, resembles a metal pipe sealed at one end; the baseplate that appears to be just a round or square piece of metal; the bipod or other barrel support structure that resembles a component of some sort of metal structure; and the sighting system. Of these four components, the sighting system would probably appear to be the most suspicious and militarylike, even to an inspector without formal military training, *if* it were visually inspected. The first three components could easily be mixed in with a shipment of odd metal parts, making their detection unlikely. Similarly, the sighting system could be intermingled within a shipment of optics and electronics. In general, when a mortar is broken down into its components and those are intermingled with physically similar items, it is very difficult to identify the individual pieces as a part of a weapon system without a visual inspection of each item.

Of all the components that comprise the complete mortar system, the most difficult to smuggle into the United States would almost certainly be the ammunition. The difficulty of disguising ammunition would be a challenge to disguise for any weapon system, not just for mortars. Ammunition has a distinctive enough appearance that an automated system—such as X-ray imaging, image interpretation, and a final physical inspection of suspicious items—becomes plausible for some sites such as ports. If a smuggling channel is indeed subject to this sort of automated inspection, any smuggled ammunition would need to be shielded or disguised. Although such disguise would require some knowledge of the screening methods used, it would not be conceptually or technically challenging. Alternatively, other smuggling channels not subject to inspection, such as small boats or private aircraft, could be used for ammunition, as they have been for drugs. An analysis of the potential of automated inspection to limit the entry of ammunition into the United States could clarify the costs and potential benefits of increasing inspections. Unfortunately, only a few advanced mortar rounds are required to enable each attack, and mortar rounds are also likely to remain relatively inexpensive. Interdiction is thus unlikely to be a robust counter to terrorist use of advanced mortars.

Finally, the precision promised by such weapons may be especially attractive to terrorists not only because it may allow them to hit their targets, but also because it means they would not have to expend many rounds to do so. Although precision is possible using man-in-the-loop systems or a terminal sensor, these options also present more difficulty to the terrorist than a GPS-only system. A man-in-the-loop system requires more training and probably a more expensive, rarer round than a GPS-only design. In particular, a laser-homing design would also require an observer with line of sight to the target, largely obviating one advantage for terrorists. Terminal homing sensors otherwise require the "right" target for homing. In contrast, a GPS-only system would offer precision enough for most targets, while requiring only the coordinates of the target. For a risk-averse group that may have only a limited number of mortars and one chance to pull off a spectacular attack, this path to precision could represent the tipping point in the decision of whether or not to attempt such an attack.

Sniper Attacks

In comparison with mortar attacks, sniper attacks that would be enabled by the new systems are simpler to describe. Modern sniper technologies significantly enable terrorist capabilities to carry out line-of-sight attacks or assassinations that would previously not have been possible, either because of their lack of skill and training or because the security perimeter around a very important person (VIP) target would be on the lookout for attacks from a shorter range. Defensive forces thus need to be aware that extremely accurate shots could come from distances far greater than those covered by customarily secured perimeters and under conditions that are not ideal for sniping (e.g., wind, nighttime).

But when is such increased range important to a terrorist group? Increased range would almost certainly be reserved only for attacks against individuals who are otherwise difficult to approach—typically, political leaders such as the President, a governor, or other nationally known officials. Other officials and all members of the public are typically accessible from much shorter range. The personal security forces needed to clear areas and to secure threatening lines of sight are not available to them, even for short ranges.

But in fact, most recent successful or attempted assassinations of public officials have been at very short range, as were John Hinckley, Jr.'s attack on President Ronald Reagan or the assassination of Yitzhak

Rabin.[11] Of course, such short-range attacks leave little chance for the escape of the terrorist, but that may not be a critical factor in planning a high-level assassination. Killing a national leader should be sufficient payoff for most groups to sacrifice a member—even in the days before the age of suicide terrorism.

Still, the existence of these sniping systems may make a difference for the personal security forces for those few individuals for whom the nation can afford real protection. Typically, this is supplied by the U.S. Secret Service to presidents, presidential candidates, and occasional visiting heads of state. Since such sniping systems are now widely available, it appears that the Secret Service will be forced to expand its secured perimeter to deny line of sight out to beyond 2 km.[12]

Attacks Using Advanced Small Arms

Advanced small arms, including not only individual rifles and grenade launchers, but also improved ammunition, short-range antitank weapons, and RPGs, could provide small groups with an important edge in combat or assaults on specific facilities or on infrastructures of concern. But there are significant restrictions on this capability as well. Overall, its appeal to terrorists will probably be limited.

Certainly this range of advanced small arms provides an edge in small-unit combat. A unit so equipped can deliver effective, short-range firepower and the availability of indirect fire with next-generation grenade technologies. It can also destroy most simple shelters intended as strong points. And the basic flak-jacket style of body armor would offer no protection to improved rifle rounds. Any or all of these capabilities could provide a decisive edge over a guard force, particularly if they were a surprise to that force. Additionally, the long-range sniping equipment described previously could also be employed in an attack.

Many sites would seem attractive to terrorist attacks if the attacks could succeed. Storage sites for nuclear weapons are one obvious exam-

---

[11] See BBC News (undated) and CNN (1995).

[12] In passing, it should be noted that such an expansion of the security perimeter would also deny laser designation for guiding smart mortar rounds, so there would be additional security benefits in countering those systems.

ple, but, not surprisingly, these are well defended. More likely targets are chemical plants with large stores of toxic chemicals or nuclear power plants with spent fuel stored on site.[13] Naturally, most such sites are guarded; the issue is the relative effectiveness of the terrorist attack.

To be effective, the attacking force would need to understand not just the operation of the weapons, but also their use in small-unit tactics. This is a significant hurdle for a terrorist group: Not only must the members master new, technical systems, but they must learn to use them together in a quasimilitary setting. Teaching small-unit tactics is complex, and capable national armies spend significant amounts of time at it. Yet without some group skills, assaulting a guarded position is fraught with difficulty.

Additionally, much of the benefit of these systems would be lost if they were *not a surprise* to the guard force. That much, at least, is under the control of the defense. Consequently, training guard forces to expect such new weapons to be used against them should be a priority.

With that change, the danger to such facilities would seem to come more from larger assault forces than from better trained and equipped ones. Terrorist groups, which typically have difficulty training individuals and lack the facilities and time for group training,[14] seem particularly likely to adopt such a strategy, should they choose to assault a protected site. Consequently, the development of these new weapons should have only a limited effect on the terrorist threat.

Attacks Using Antitank Guided Weapons

Although the advantages of ATGW—their portability, excellent accuracy, powerful warheads, ranges of up to four kilometers—are clear, there has, so far, been little terrorist use of these weapons. During the 1990s, the Hizballah group in Lebanon used a small number of ATGW against Israeli armor. The Iraqi armed forces had several types of ATGW available but used few of those in the 2003 war. There is little if any evidence that insurgent groups in Iraq have employed these weapons, despite the fact that some number of those weapons could

---

[13] National Research Council (2006) covers the latter attack's potential consequences.

[14] Forest (2005). See also Teaching Terror (undated).

easily have been obtained from looted Iraqi Army and Republican Guard stocks.

Indeed, Iraq's arsenal included the old Soviet-made AT-3 *Sagger*, the much more advanced Soviet AT-4 *Spigot*, the European-made *Milan*, and the heavy European *HOT* missiles. The lack of insurgent use of this type of weapon may be due to several factors, including few weapons being in their hands; insufficient training (although some members of the insurgency in Iraq are clearly former soldiers, some of whom could be familiar with these systems); or maintenance issues of the weapons (most or all of them may be in poor condition). The virtual absence of insurgent use is certainly not due to a lack of targets or opportunity to use the weapons.

In the future, insurgents or terrorists could employ ATGW in the United States. As mentioned above, the tactical advantages of the weapons are clear. An ATGW is designed to destroy a tank. Consequently, an attack by a modern ATGW would demolish all civilian vehicles, including those used by governmental officials. Armoring such vehicles is not an available option.

Two drawbacks to these weapons are clear:

- difficulty in conducting training (even though computer-assisted training devices and simulators are available for ATGW, the operators still have to undergo considerable training to properly use the systems)
- perceptions on the part of terrorists that they can achieve their purposes with less sophisticated systems (e.g., car bombs or mortars).

Unfortunately, effective terrorist attacks do not require the latest, fire-and-forget systems. Older, man-in-the-loop systems would be just as effective. Newer systems are easier to use, but not dramatically so. The drawbacks above would largely remain, with only the training needs being reduced (but not eliminated). Since the older systems are already widely proliferated, this type of terrorist attack appears to be one that we cannot meaningfully limit through technical or additional procedural controls.

Attacks Using Limpet Mines

The mode of attack used during World War II is very similar to how limpet mines would be employed today. Underwater swimmers would attempt to reach their targets undetected to attack critical, symbolic, or otherwise attractive shipping targets. What would be the most lucrative targets for terrorist-emplaced limpet mines? Probably passenger ships such as cruise liners or passenger ferries, which can carry many people, or vehicle ferries, and which can quickly sink if the large vehicle space fills with water. In particular, either cruise liners or long-distance ferries would allow a timed detonation to damage or sink a ship far from shore. Although the sinking of an oil tanker or large cargo ship would be a serious event, it would not compare to the loss of hundreds of lives on a cruise ship. For this reason, certain types of vessels—cruise ships, large ferries, liquid natural gas ships, and other vessels carrying hazardous cargo—would be the most appropriate ships for protective actions and countermeasures. As mentioned above, modern limpet mines employ increasingly sophisticated antitamper devices. Therefore, appropriate safeguards would have to be taken to ensure that a mine discovered on a ship not be removed in a way that results in a detonation.

Comparing the Most Threatening Advanced Conventional Weapons

The five classes of advanced conventional weapons discussed above—advanced mortar systems, sniper weapons, advanced small arms, ATGW, and limpet mines—present different problems for the United States and require different countermeasures. The threat from advanced mortar systems, particularly when GPS-guided rounds become available, is by far the most worrisome. In the hands of terrorist groups, these advanced mortars would enable qualitatively new attacks, such as assassinations within secured perimeters. More important, the use of these weapons seems difficult to counter through training or technical enhancements to security forces. The range and thus available area from which attacks can be staged is simply too large. This leads us to consider technical and other control measures on the weapons, particularly for these mortar systems, in the next chapter of this monograph.

In contrast, the other four categories discussed above—sniper weapons, advanced small arms, ATGW, and limpet mines—would enable a smaller qualitative change in potential terrorist attacks. To be sure, security forces need to be aware of the potential of these systems in order to counter them. Otherwise, if a new capability is unexpected, the first use of these systems could still produce a dramatic result, such as the assassination of a leader at an unexpectedly long range, the quick suppression of a guard force, or the rapid sinking of a large ship. Although the consequences of such a successful attack might appear game changing, it seems possible to largely reduce the likelihood of even an initial success by making changes in the training and procedures of security forces, as discussed above. All four of these categories of advanced conventional weapons require such changes; fortunately, the cost of such changes does not appear to be large.

CHAPTER FOUR

What Opportunities Exist for Controlling Weapons of Particular Concern?

Given the potential hazard of some advanced conventional weapons should they be diverted into terrorists' hands, mechanisms for denying or limiting the utility of such weapons are clearly desirable if they do not interfere excessively with the military utility of advanced conventional weapons. The established mechanisms for controlling the use of weapons by unauthorized users are known by the term *use controls*. Use controls can be thought of as a collection of procedural, technical, and policy tools that arrayed together can limit access to, and ability to use, advanced conventional weapons. Some controls are focused on preventing the underlying technology from being transferred, e.g., the Missile Technology Control Regime (MTCR). Other methods focus on limiting the lifetime of weapons (potential MANPADS control) and others focus on limiting the ability to use a weapon (the best example of this being the introduction of permissive action links to the nuclear forces). In all cases, the key to effective use of these control mechanisms is their early application before the technology becomes widespread and effectively uncontrollable. To understand use control, it is necessary to understand what exactly we are talking about in terms of use controls, where they might be applicable, and how they work in practice.

Types of Use Controls

The term *use control* captures a large number of very different approaches for limiting the ability of unauthorized personnel to gain access to, or to use weapons. The three major categories of control are thought to exist in very different domains. First, there are policies that affect how weapons and technologies are handled in a macro sense, such as whether technologies can legally be pursued or transferred between states. Next are procedural controls that affect the day-to-day handling of weapons by authorized users. Finally, there are technical controls that focus on the weapons themselves and seek to limit their utility in the hands of enemies from the standpoint of either actually using them or gaining access to sensitive technologies and materials. Each of these approaches has a limited domain of utility and have, at least in the United States, been handled by different parts of the government. For instance, both the State and Commerce departments are at the forefront of managing many counterproliferation issues, while the Department of Defense and military services are responsible for both day-to-day procedures for handling weapons and for any technical controls that are built into weapon systems.[1]

One aspect of use controls needs to be understood. Use controls are not simply targeted against some particular weapon; they are targeted against that particular weapon at particular stages of that weapon's life cycle. At other stages, they may be ineffectual, or entirely absent. Consider a life cycle of a weapon assigned to fielded forces. Figure 4.1 illustrates a simple, seven-phase life cycle (manufacturing, main storage, depot-level maintenance, operational storage, forward deployed storage, operational usage, demilitarization), with three distinct states of weapon readiness and three levels of risk associated with diversion at the various locations and in transit. The risks range from low to intermediate to high. A weapon starts its life in a manufacturing facility and ends its life at a demilitarization facility, many times with the launcher and control portions separated from the destructive payload. In some cases, portions of a weapon can even be recycled into newer weapons, creating a flow *from* demilitarization. In general, weapons move from manufacturing facilities to large main storage locations, where usually all the components of a full-up system

---

[1]   There is an important exception for nuclear warheads. The U.S. Department of Energy is responsible for technical measures protecting the warhead portion of the system.

Figure 4.1
Exemplar Life Cycle of Advanced Munitions

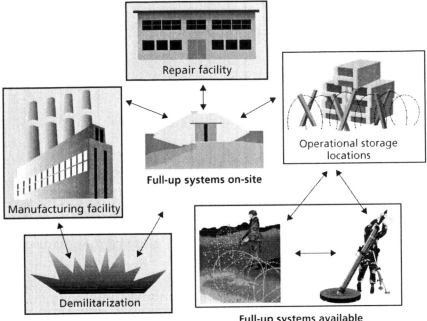

RAND MG510-4.1

could be expected to be found, but not necessarily located in close proximity to each other, producing an intermediate risk.[2] A weapon system might move to an operational or forward deployed storage location in what amounts to a fully capable status. Finally, systems actually in the field for use must be fully capable. In these three stages, fully capable systems are likely to be available for immediate use and have the greatest potential for diversion. The risk associated with a single diversion event increases as the weapon moves from an unready state with components possibly dispersed to a ready state with all components present and moves from an environment with extensive physical security and procedural control to a fluid, operational environment.

---

[2] Explosives' safety requirements usually require substantial separation of the explosive portion of the system from other activities in which people could be exposed to danger, but they do not require separation from nonexplosive elements of the weapon system.

The types of use control most relevant vary considerably in a weapon's life cycle. Figure 4.2 shows the relevant domains of different control regimes. Policy and procedural controls are focused primarily early and late in the weapon's life cycle. Limiting who has access to the weapons and how the weapons should be handled matters a lot in terms of security and has minimal operational impacts. In the field, operational necessity dominates, and the other hazards are regarded as acceptable trade-offs, given the realities of the battlefield. Consequently, once the weapons move close to operational use, procedural and policy controls need to be supplemented with technical approaches to guard against weapon use. The reasons for this change are simple. Procedural controls almost by definition hinder operations, and they have a smaller effectiveness because all the parts of the weapons need to be present and ready for immediate operational use (denying one of the more powerful procedural controls of separating key system elements when possible).

With the general framework in mind as to where each of the use control approaches might be focused within a weapon system's life cycle, it is now possible to walk through a brief discussion of each main approach. This slightly more detailed discussion will highlight the main limits of each approach and demonstrate how, through a careful structuring of combinations of control mechanisms, it may be possible to effectively implement a use control approach for advanced conventional weapon systems without excessive costs or operational complexity.

Policy and Procedural Controls

Policy and procedural controls are two related but not identical systems that use rules of behavior to control the technology. In one case, the rules might be formalized in bilateral, state-to-state agreements or in a multilateral international treaty. In another, the rules might simply consist of a less formal agreement that guides the behavior of the parties. In all these cases, the controls are based primarily on an agreement of the parties involved to play by the rules. State-to-state rules might include which countries have access to technologies and weapons, whether the weapon technology can be exported legally to third-party

**Figure 4.2
Use Control Application Domains**

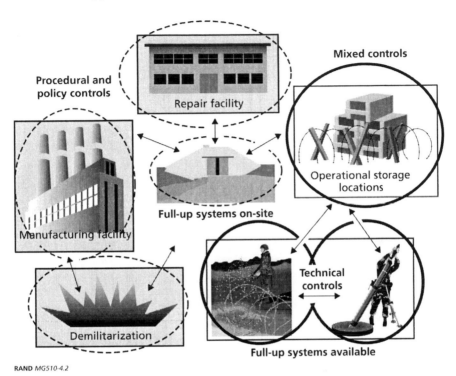

RAND MG510-4.2

entities, and what sanctions are associated with noncompliance. And in all these cases, national laws and regulations are needed to enforce the agreed upon separation of sensitive materials, or the use of guards and locks as part of normal military operations.

Typically the rules controlling state-to-state behavior are governed by agreements such as the Wassenaar Arrangement or the MTCR that specify the rules of the road and identify which systems are captured by the agreement. These kinds of agreements govern many aspects of state-to-state interaction, particularly in regard to classical proliferation issues involving nuclear, chemical, and biological weapons, as well as arms transfers. These agreements are focused on keeping weapons out of the hands of state entities that are deemed unreliable, and on decreasing the general availability of underlying technology.

These institutional agreements are by nature difficult to implement, slow to change, and require consensus to have any utility in practice. Furthermore, the rules implemented by the formal regimes often preserve sufficient room to maneuver, allowing member states to pursue their own objectives at the expense of the collective interest of member parties. For instance, range limitations of cruise missiles are a focus within the MTCR, but they can easily be circumvented. Arguably, such limits have little practical meaning. In practice, this means that such agreements exist to codify the already extant behaviors of member parties and will tend to be something that can only be useful if there is a consensus such that the objectives and approaches of the agreement are already internalized into each member state's national policy. Without strong support from such an internalized consensus, the natural tendency to preserve operational flexibility and weapon reliability can reduce controls. Additionally, nationalistic concerns about ceding sovereignty or prerogatives to an international agreement or group can further erode support for a control regime. Thus, without internal support, controls would tend to degrade to a simple attestation of having met security requirements, with minimal or nonexistent inspection procedures for all but the most dangerous of weapons (where presumably some national support would exist).

Procedural controls, such as requirements to handle weapons in particular ways, are routine in militaries of the world. Safety and operational requirements routinely dictate environmental and security measures associated with weapons. For instance, the requirement for locks on magazines is routine, not just to prevent theft of weaponry but also to prevent accidents. However, the story differs substantially in terms of the ability to monitor diversions of the weapons or to withstand assaults directed against magazines or storage areas. In this case, the cost is not simply a lock (which may or may not actually be used); it is for personnel associated with the protection of the facilities or changes in how a deployed force should operate. In both of these instances, resistance to guidance can be expected to be the norm, if only for normal bureaucratic reasons.

The problem is even more serious when considering the possibility that weapon diversion could occur at any point in the weapon's life cycle. Although procedural control would address many of the issues, for example, rules minimizing the colocation of munitions and fire control elements, there are major issues associated with implementation. For instance, some elements would be governed by defensewide or service regulations; others would be associated with industrial controls; and still others with requirements for safety and security of dual-use components, governed by civil agencies. In principle, this shared responsibility can work, but seams will develop as responsibility is spread and as different interpretations of the "requirements" for control develop. It would appear that the ability of procedural controls to avoid leakage of small numbers of weapons and critical technologies under all but the most stringent controls are probably fairly limited. Past U.S. experience with munitions controls, especially of items transferred to second parties, makes it unlikely that this will provide sufficient protection against diversion.

Before leaving the discussion of control regimes, it is worth noting that the existing international control regimes have been primarily oriented toward preventing certain countries from acquiring militarily significant numbers of advanced weapons or gaining access to production technologies to produce their own version of the weapons with minimal new engineering and production requirements. However, much smaller numbers of weapons have a greater meaning for homeland security. Theft of weapons, especially easily transportable weapons, are of great concern under most circumstances, and indeed, as shown earlier, even the availability of a modest number of advanced weapons can pose some difficult problems for those charged with protecting high-value physical targets and personnel.

Technical Controls

Technical control regimes focus on making the weapon difficult for the enemy to use and, to a lesser extent, preventing exploitation of sensitive elements of the weapons. In the world of conventional weapons, this is a new concept. Designing robust control regimes has been perceived as difficult: They add to operational complexity and cost and were thought to be of insufficient importance in terms of military utility to warrant inclusion. The only areas in which these types of controls have existed have been associated with nuclear weapons, some command and control elements, and to a lesser extent, sensitive communication and intelligence equipment. The latter cases were primarily associated with denial of access to sensitive elements of the system. Consequently, when it comes to most weapon systems of relevance to this study, this is a new concept.

The closest analog to what might be required for use control on advanced conventional weapons comes from the nuclear weapons community development of permissive action link (PAL) technology. A PAL is

> a device included in or attached to a nuclear weapon system to preclude arming and/or launching until the insertion of a pre-scribed discrete code or combination. It may include equipment and cabling external to the weapon or weapon system to activate components within the weapon or weapon system.[3]

PALs facilitate extremely tight control over the use of nuclear weapons by making them unusable without the proper codes. PALs could also, if appropriately engineered, support selective arming of some or all weapons, as well as allowing for the disarming or the permanent disabling of weapons. Furthermore, an appropriately designed PAL system would reflect the idea that compromise of one weapon should not endanger the security of other protected weapons, forcing attacks on the PAL system to be one weapon at a time. In all these cases, the key element of the system is a carefully engineered set of mechanisms designed to operate with high assurance even if the weapon itself has somehow fallen into unauthorized hands and is subject to physical attacks.

For advanced conventional weapons, the main challenge is building a system with some of the characteristics of the PAL system for nuclear weapons, without resorting to the same high degree of design and manufacturing costs, operational complexity, or security models associated with the command and control of nuclear forces, both stra-

---

[3]   See U.S. Joint Chiefs of Staff (2001 [2007]). For general descriptions of the function of PALs and how they may work within the nuclear weapon community, see Blair (1993), Hansen (1988), and Cotter (1987).

tegic and tactical. Ross Anderson provides a valuable discussion of the applicability of lessons from the nuclear command and control community in which he highlights the importance of the authentication mechanism (determining whether a command is not only correct, but properly authorized), shared control schemes, and the need for tamper resistance.[4] The lessons are important, since they strongly suggest the major lessons to carry away from the nuclear community: Security needs to be an important element of design, and implementation details will determine the success or failure of any security scheme.

The tensions between usability and security are very obvious in the conventional weapons case. From a military standpoint, the most desirable situation is for a weapon to operate properly in the field regardless of which user is controlling the weapon and for targets to be struck effectively and rapidly under all conditions. Security, on the other hand, would like to deny use of weapons to all unauthorized users under all circumstances and at the limit would like to have sufficient specificity to control individual weapon use. Obviously, a balance needs to be struck between usability and security, with operational risk being traded off against security of the weapons and the dangers they pose when they fall into the wrong hands.

A key insight into striking a balance between security and usability is the understanding that there are two very different kinds of risks being managed. The first-order military risk is that the weapon will not be available when needed by military users. The second risk, primarily borne by nonmilitary entities, is that the weapons will be diverted and pose both an immediate and enduring threat. With this insight, it is possible to consider an ideal use control framework—one that strikes a balance between these two types of risks in setting the performance objectives for a use control system.

On the battlefield, military users do not see it as practical to go through elaborate authorization procedures before permitting any individual weapon use. From the military's vantage point, it would seem

---

[4] For a description of general security approaches, see Anderson (2001). See also Schneier (1996) for a discussion of some issues associated with cryptographic authentication schemes. Also see Anderson et al. (2005).

desirable to enable devices for an extended period of time to increase usability by the troops, albeit at the expense of some enhanced short-term risk from diversion of the weapons on the battlefield. On the other side of the issue would be those concerned about extended windows of vulnerability, which would focus on nonbattlefield security issues. From their vantage point, it would be helpful if the weapons were relatively inert whenever not in possession of an authorized user or perhaps when not in an authorized location. Also, there might be some significant interest in making sure that compromise of control on one system will not lead directly to compromise of control on other weapons.

Striking a proper balance between these two set design objectives can be satisfied several different ways, but the basic design criteria are clear: Military users just want the systems to work, and those looking for extended controls want them to be safe whenever they are not in the hands of the authorized users. How can such a scheme be technically accomplished? It would seem that a variant of the approach discussed in the PAL example would be appropriate, with the modification that an enabled system would be fully operable only within a particular time window or geographic box and not fully capable in other periods or places.

There are several parts of the use control mechanism that are important: a tamper-resistant locking device linked into a critical function of the device, a key to unlock the device, and a mechanism for getting the authorization codes and keying material to the device. An additional demand exists for any system with an expiration timer—the ability to get access to it at the proper time. The actual engineering and the system device are well beyond what can be properly addressed in this monograph, but we can outline an analytic, conceptual design that suggests the technical characteristics of a system to meet the demands of the two communities.

A Conceptual Design of a Use Control System

The first issues to consider are the identity and level of sophistication of the possible attacker and then what is expected from the system. Why consider the attacker? Simply put, the level of protection must be commensurate with the opponent's capabilities to defeat the protections and the tools available to the opponent. Potential attackers run the gamut from unskilled individuals, skilled individuals, groups of varying skills, and corporations, all the way to nation-states. This list illustrates this range of potential attackers:

- nation-states with advanced indigenous capabilities
- nation-states with foreign industrial support
- corporate entities
- subnational groups with limited support from one of the above
- subnational groups with no outside institutional support
- individuals with advanced technologies.

An individual can be very sophisticated but will tend to be constrained in terms of resources available, especially technical resources. For instance, an attacker might have to resort to nondestructive attacks to recover information, whereas a company or state may have access to advanced tools, such as electron microscopes that can help read out information from hardware and firmware to assist in an attack. Overdesigning the use control system will lead to excessive cost and complexity, so making appropriate design decisions is very important. For instance, if the United States is concerned about what skilled individuals might do, the additional levels of protection designed to counter sophisticated attacks by a national laboratory would be unnecessary. To provide a context for choosing the level of protection, it helps to first examine the various reasons that we might want to exercise controls.

There might be instances in which higher levels of protection are desired for the system or for a component of a system, such as the following:

- Prevent adversaries from gaining access to full capabilities of the weapons.

- Prevent adversaries from any destructive use of the weapon.
- Prevent adversaries from copying the system.
- Prevent adversaries from exploiting the system to prevent effective use of the system by friendly forces.

Preventing use is only one possible objective for controls. Interestingly, it is probably also the simplest, inasmuch as all of the others either prevent exploitation of the system or interfere with fairly basic functions such as warhead functionality. Prevention of exploitation requires "denial" systems that destroy key information and components to prevent compromise of technologies even if a weapon or system must be sacrificed in the process. This is an extremely demanding objective and requires special purpose devices.

It would seem that, for the conventional weapons we have been considering, the appropriate thresholds are fairly clear. Designing toward the low end would be consistent with the level of homeland security concerns about these weapons altering the balance between attackers and defenders. A focus on countering subnational groups and individuals with limited support and on preventing an adversary from gaining full access to the system would render the new generation of weapon no more (nor less) dangerous than current systems. This could be made more robust against more sophisticated attacks against the system by government-supported entities, but, in all likelihood, any group able to exploit the system would probably already have most of the capabilities to make a similar system.[5]

The design requirements for our suggested control system would allow the weapon to be in two states: disabled or enabled. Once enabled, the system is to remain in that state until a defined time occurs, and then the weapon returns to a disabled state. For weapons with a

---

[5] There are some reasonable arguments for working toward higher-end protections to add one more barrier to copycat systems. However, that issue needs to be addressed in the context of how the system will be physically packaged. For instance, some tamper resistance may be built into the system to protect the most sensitive cryptological elements and, depending on whether integration of other subsystems into the security model makes sense, may include some other sensitive elements to make altering the system more difficult. However, completely preventing reverse engineering of the system can be extremely difficult.

reliable and secure clock, such a system would appear straightforward to implement once the system has been enabled after being moved to forward locations. Similarly, for weapons with a secure knowledge of position, geographic regions can define the limits of allowable use. In practice, both time and position are available for systems using GPS or a similar satellite system.

Secure equipment (that could verify authorization codes or keys against master databases) would be used to activate groups of weapons, and secure timers would be started for anything that can remain powered up for extended periods. For added security, software could be zeroed while the weapons are in the disabled state, except for a primitive loader for software. Key revocation would have to be accommodated should a root key be compromised. In this model, the weapons and enabling devices must establish trust between one another, but only the enabling device needs to ensure the validity of messages received from higher-level authorities.

The story gets a little more complicated for any nonmonolithic system such as a GPS-guided mortar system, in which one part of the system, the mortar round, is nominally a "wooden round"[6] that "wakes up" only immediately before use.[7] In this model, the weapon has no state information when it awakes; it needs to gather this information from external sources. It needs to "handshake" with the part of the weapon system that will authorize it (in this case, the launcher) and then use the information to enable a critical function. For GPS-guided weapons, the key functionality is not warhead arming; it is the guidance system. This means that the system can be designed to operate in a ballistic mode—that of an ordinary mortar when not authorized—and will guide correctly only when properly authorized to use the GPS signal. A key element in this system would be a common shared secret—the ability to decrypt the coded GPS signal—and this

---

[6] This is in contrast to a system that is on and monitoring itself on an ongoing basis.

[7] The system may not be a true wooden round. One of the things it may have is a whitelist or blacklist of keys loaded at the factory or in depot-level maintenance. Also, the ability to have the system come out of deep "sleep" for maintenance may be a better match than a true wooden round.

ability would be used to establish trust between critical elements of the system.

In general, the system would be expected to be in the disabled mode and readied for combat or training only when moved to a forward location. The system would have extended periods of time (and perhaps space) in which all elements are capable, such as allowing unrestricted typical military operations there. After a preset authorization period, the system would have to be reenabled. Disabling could consist of simple remote rekeying of the device or an erasing of the device to require more extensive reloading and reauthorization.

The actual engineering of a system can be quite complicated considering the interrelationship between elements involving cryptographic systems, GPS receivers capable of decoding the secure GPS signal, and an integrated GPS/INS (inertial navigation system) for weapon guidance. In addition, how the key management issues are handled is very important in accomplishing the overall objective of limiting diversion. Fortunately, it appears, at first blush, possible to engineer some system that can make use of existing elements to accomplish the goals. However, an added wrinkle is that the system needs to be designed not only to be resistant to an entity posing as an authorized user; it also has to deal with the possibility that a critical element of the system itself has been altered. Why is this important? Because the system needs to protect against a simple hardware hack that either bypasses the security protocols and systems or alternatively replaces everything, including the embedded security system. Fortunately, the use of a tightly integrated GPS/INS/crypto system means that simple attacks would not likely prove effective, since an appropriate design could force replacement of not only the crypto elements, but the entire integrated GPS/INS as well as the flight software for guiding the weapon. Tight integration of components, as well as use of secure systems, might prove especially useful, both in increasing the difficulty to the attacker and in posing a problem for the would-be attacker.[8]

---

[8]   The idea of raising the bar for the attacker is an important difference here, relative to the nuclear arena. Here, hardware attack on the system will only be capable of compromising a single system if the rest of the security model has been properly implemented. In the nuclear

At this point in time, it is not clear whether solutions can be built using existing components or whether new components would be required to actually implement a viable system. Component elements such as the secure SAASM (Selective Availability Anti-Spoofing Module), which provides a secure, tamper-resistant GPS capable of decoding the P(Y)-code GPS signals of the Precision Positioning System, may provide a useful starting point for a design of this tightly integrated system, by virtue of both its security and the criticality of the GPS information for its mission success. It may require the additional systems built into the navigator itself to guard against functional removal of a SAASM receiver from the system and the replacement of it with another module. Such an approach can be seen in the Trusted Computer Base reference design.[9] A key element of the system is a carefully crafted set of hardware- and software-based validation activities to protect against alteration of the key hardware and software elements by malicious terrorist actors. If a security-focused design is included, in which the navigation element of the system protected itself from alteration of the GPS signal-processing element of the system, as well as validating authenticity of the authorization, it may be possible to greatly limit the possibility of a malicious terrorist actor using the system.

Whatever the approach taken, it is important that the security-related elements be built into the system from an early point. A carefully thought-out design, one that takes into account as wide a range of potential attacks as possible, is needed to guard against simple alteration of the system. For instance, skilled individuals can replace many parts of modern systems if they are willing to spend the time and effort to understand key elements and if the functional elements have well-defined and predictable interfaces and behaviors. Designs that depend purely on obscurity or that make too many assumptions about the level of threat are almost always subject to an effective attack.

A good example of attacks on systems can been seen in efforts directed at breaking the security of "smart" cards, crypto proces-

---

arena, PALs are intended to guard against the compromise of a single weapon to the greatest extent possible.

[9]   See Trusted Computing Group (2004).

sors, and associated security systems by Anderson and by the efforts of Huang and others in the community to attack the security of the Xbox®.[10] In both cases, the attacks are based on extracting secrets from the hardware to obtain a key to allow further operations of the system. The university-based groups are at the high end of the attack spectrum about which we are concerned, while Huang represents the attacks of skilled individuals with some community support. The main differences between the individual and group attacks are the breadth of technical knowledge available on site and the kinds of specialized equipment used in the attacks.[11] In both cases, very sophisticated attacks were used to probe systems to allow the attacker to impersonate the authorized user of the system or to extract critical secrets of the system to allow other attacks in the future. It is clear that failures need not rely upon some mistake associated with the underlying crypto systems and key management schemes; instead, attacks on the hardware itself can yield significant returns if design or implementation errors creep in at any stage. For instance, the attack on the Xbox was successful in part because of the assumption that a high-speed bus between two system elements could not readily be monitored, which helped extract otherwise protected information.

Combining Policy, Procedures, and Technical Solutions

The soft protection schemes provided by policy and procedural control of systems have important limitations in the types of protection they provide. Although they do serve to decrease the chances of a fully

---

[10] Huang (2002, 2003). The combination of these sources provides an excellent example of a well-documented effort to understand a protected system using a variety of different techniques. Of particular importance were the ability to tap into the hardware to extract hidden information and the advantage the attacker had because of Microsoft's particular design choices. In contrast, Anderson's group has focused on attacking the security processors, and that work is documented in Bond and Anderson (2001). Xbox® is a registered trademark of the Microsoft Corporation.

[11] Individuals can be greatly augmented by help from the outside. For instance, Huang (2003) reports having been greatly assisted by outsiders interested in hacking the Xbox.

capable weapon being diverted, they do not protect against the impact of weapons being diverted. In contrast, a physical use control system makes it difficult to use a weapon once diverted. By combining the two elements together in an overall approach, it should be possible to make both the likelihood of diversion and the consequences (should a diversion still occur) much smaller than they would be otherwise. For the most part, the procedural controls would seem to be focused where they would have minimal operational impacts to legitimate users—away from the battlefield. Technical solutions would be focused only at those places where all the system elements would routinely be expected to be together and would depend on a design that links to some key operational characteristic of a system, such as the dependency on GPS; creates a robust and secure system for enabling weapons; and defines limited and selective windows of full operational capability. In concert, these limitations would decrease the chance that any diverted system would present a prolonged threat from terrorist use.

Implementing this process in the real world will be a challenge. The technical systems are not common to most battlefield weapons, and they will certainly cost more in design and development than a less secure design would. Furthermore, by its very nature, the system will be viewed as less trustworthy on the battlefield because it is designed to be nonoperational should enabling errors be made. In the realm of policy and procedures, there is a similar story. Limitations impose some burdens, both financial and operational. If these weapons are handled differently from all other weapons, there will be greater costs, operational complexity in the logistics systems, and a general concern that these kinds of models would be extended to other systems that would again drive up the costs and complexities of the logistics system.

Almost certainly, technical controls will only be practical for some new systems that have not yet been fielded. For such new weaponry, though, there is the potential to combine both technical and procedural controls, significantly reducing the likelihood that terrorist groups will gain access to the new capability. But the quite visible costs of implementing these controls will only be accepted if the threat from terrorist use is both high and widely understood and if the controls are seen as effective.

CHAPTER FIVE

How Might Use Controls Deter Terrorist Organizations?

Whether or not a terrorist group pursues a new, advanced weapon can be framed as a judgment about costs and benefits. In this context, the choice to seek a specific weapon will depend on the terrorist group's assessment of how potential benefits compare with the costs of obtaining a weapon and on how the apparent costs and benefits of that advanced weapon compare with other tactical and technological options available to the group. This calculus may be an implicit rather than an explicit process, and decisions may be based on cost and benefit criteria that are idiosyncratic to the terrorist group. Nevertheless, a process with these basic components will underlie decisionmaking at the individual and organizational levels.

Such cost-benefit decisions are further complicated by uncertainty. Depending on the information available to the terrorist group at the time, it will face two different, but complementary risks. They are

- the risk that the group's cost-benefit judgments about the technology are incorrect and it is choosing to adopt a weapon that is not, in fact, supportive of its objectives
- the risk that the group's attempt to adopt the technology will fail and it will pay the costs associated with doing so without gaining the desired benefits.

An organization can reduce these risks by seeking out more information and expertise before committing itself to an adoption effort. More information can provide more certainty about the technology's actual costs and benefits, thereby enabling a better decision. Other types of information can also reduce the risk that the attempt to adopt will fail. Gathering such information takes time and effort, however, which increases the total cost of the technology to the group and delays any adoption activities. A group's judgment of whether to pursue a technology at a given time will therefore be based on whether it will provide a sufficient net benefit, how certain it must be of that benefit, and how much risk of failure it is willing to take in the adoption effort.[1]

The effort to change terrorist decisionmaking about advanced conventional weapons can be viewed as targeting groups' perceptions about the costs, benefits, and risks of acquiring and attempting to use the weapons. As a result, assessing the utility of technical approaches or other control regimes requires a framework for breaking down the elements of that decisionmaking process to identify the forces that shape it and how they might be influenced.

The Benefits, Costs, and Risks of Adopting Advanced Conventional Weapons

Studies of a variety of organizations and their technology acquisition behaviors have identified a range of factors, as shown in Table 5.1, that can positively or negatively influence an organization's judgment about the costs and benefits of a technology and help explain the adoption decisions made by different organizations.[2] Such a taxonomy provides a structured way to examine the factors that influence the perceptions of terrorist groups of the costs, benefits, and risks of adopting a new

---

[1]   Rogers (1995), p. 14.

[2]   For further discussion, see Baptista (1999); see also Rogers (1995). In most cases, the decision to adopt or not adopt a technology is more accurately characterized as the decision to adopt or *defer* the adoption decision, since the group could choose to pursue a "rejected" technology later. As a result, these factors—shaped by any new information the group gains since its initial encounter with the technology—will also apply to any subsequent instances when the group might revisit or revise its initial judgment. Whether or not a group will change its decision later will be affected by the openness of the group and its decisionmakers to new information and a willingness to revisit their initial judgments.

Table 5.1
Factors Affecting Technology Adoption Decisions

Category	Attribute	Adoption Decision Factor
Technology	Comparative advantage	Technologies with larger apparent advantages compared to currently available options will be more readily adopted.
	Compatibility	Technologies that appear compatible with the current ways in which the organization operates will be easier to adopt and, therefore, less risky.
	Complexity	How simple or complex a technology appears affects perceptions of how risky it will be to adopt.
	Trialability or observability	"Test-driving" a technology before committing to adopt can provide significant information and reduce adoption risks. Although inferior to trying the technology itself, observing its use can provide information to reduce adoption risk.
	Price	The more expensive a technology is to a group, the higher the stakes in deciding to adopt it.
Group and its social systems	Internal group decision structures	Depending on the authority and other structures within the group, adoption decisions could be made collectively or individually. The nature of these internal structures could affect when and how a group decides to pursue a new technology.
	Communication channels	A group's ability to gather additional information to inform its adoption decision and reduce the inherent risks involved depend on the nature of the communication channels available to it.
	External environment	Activities by organizations or individuals outside the group can affect the adoption decision. For example, external proponents of change seeking to "sell" a group on a specific technology could contribute to its adoption decision. More generally, the spread of a technology among other groups could provide a less focused, but still relevant, pressure on a group to adopt it.

SOURCES: Adapted from Baptista (1999) and Rogers (1995).

weapon. The assessment of any control measure requires an understanding of how the natures of different terrorist organizations may affect the relative importance of different factors.

Benefits Drive the Adoption Process

In most cases, the potential benefits of new technologies are the initial driver for adoption. The scenarios described in the previous chapters have graphically demonstrated the potential tactical and operational advantages that next-generation conventional weapons could provide to terrorist organizations: Such weapons can provide entirely new capabilities and options to these groups, allowing them to carry out current operations more effectively, safely, or reliably. Possession and use of advanced weapons can similarly provide groups with "image" benefits as well—use of weapons associated with professional militaries can provide prestige to a group that can, in turn, bolster recruiting and fundraising and generally advance other group goals.[3] This prestige factor can become a potentially powerful incentive for terrorists to covet advanced weapons and seek them out. J. Bowyer Bell, a scholar who spent many years studying revolutionary and terrorist movements, observed the following in 1987:

> Every revolutionary movement, almost without exception, avidly wants new, elegant weapons. . . . And it is abundantly clear that in the immediate future the world will be filled with those engaged in armed struggles. . . . Certainly there is no doubt that frantic men will seek any and every means to advance their cause.[4]

The potential benefits of a new weapon technology to a terrorist group can go well beyond simply the desire for something new that can enable different tactics and operations. The perceived benefit of advanced systems to a group will also be driven by the actions that have been taken against it and how those actions affect the group's current capabilities. If security measures that have been taken or shifts in the political environment make a group's current weapons less effective (e.g., a particular hardening measure that makes conventional explosives less useful), perceived benefits of advanced weapons that can

---

[3]   See, for example, discussion of RPG use by PIRA in Jackson (2005).

[4]   Bell (1987), p. 50.

neutralize that countermeasure increase.[5] Such shifts in environment transform the desire for new weapons from a "want" (as described by Bell) to a "need" for maintaining the group's capabilities.[6]

But the benefits of a new weapon can become a driver for adoption only if the terrorist group knows that the benefits exist. Because terrorist organizations are clandestine and frequently lack the opportunity to experiment with a broad range of different weapon technologies, there is no guarantee that a specific group will be aware of the full potential and applicability of particular weapon technologies. In general, a lack of familiarity with a technology can be a major barrier to "first use" by an organization, since the lack of awareness and skills can reduce the perceived benefits of pursuing the technology. However, this barrier can be diminished if a group can observe the technology in use, and it can be completely removed if a group can actually experience the use of similar or related technologies. For example, the widespread use of mortars by a number of groups in the Iraq insurgency is of particular concern. If and when individuals involved in the insurgency leave Iraq and pursue violent activities elsewhere, that experience will likely strengthen their view of the benefits of similar, more advanced technologies.

Cost Considerations

Although the benefits of an advanced weapon create an incentive for adoption, terrorist groups—like individuals—do not necessarily *acquire* every new technology they might desire. Focusing only on the potential benefits of a weapon, therefore, risks reaching erroneous conclusions, especially since the costs of pursing new technologies will partially (or, in some cases, fully) offset its benefits. The observation that most terrorist organizations are operationally conservative[7] in their weapon acquisition strategies—relying primarily on explosives and

---

[5] See discussion in the companion volume, Jackson, Chalk, et al., (2007).

[6] The fact that familiar weapons—"guns and bombs" as described previously by Hoffman—frequently continue to serve effectively to achieve many terrorists' operational goals is likely a key reason for the comparatively sparse use of advanced weapons by such groups.

[7] Hoffman (2000).

firearms—likely results from the costs of many advanced technologies outweighing the benefit of utilizing tried and true attack modes.

When adopting a new weapon, a terrorist organization may face a range of costs. Drawing on the taxonomy in Table 5.1, it becomes clear that costs can arise from incompatibilities in how the weapon meshes with the group's activities; difficulties in using it well if it is more complex than weapons currently in use; inherent financial, effort, training, and other costs involved in acquiring it; and any detrimental changes in the group's external environment that occur as a result of adoption (e.g., reactions by the group's rivals or counterterrorist organizations). Depending on the group's specific circumstances, these costs may make pursuit of specific weapons appear unwise.

A variety of factors on the supply side for weapons can affect the perceived costs of a new technology and make it more attractive to terrorist organizations. If external circumstances present a group with an attractive opportunity to gain a new technology, the organization need not expend time, effort, or other resources to seek it out for itself. Such opportunities, often provided by actors such as states or sympathetic groups in the organization's external environment, may include the chance for members to try out the new weapon and to ensure that it is compatible with the group's tactics and way of operating, thereby allowing the group more complete information on the costs[8] and on any adaptation that might be required for the group to actually use the weapon effectively. As a result, in some contexts, terrorist organizations may opportunistically pursue weapon systems that they would otherwise not have pursued if they had to bear the full costs of doing so on their own.

Risks in Acquiring New Technologies

Because of the pressure to appear successful and effective, terrorist groups have been described as risk averse in a number of important ways. Although they take violent actions that involve taking specific

---

[8] Such opportunities provide a similar chance for the group to better understand the likely benefits of adopting the technology, simultaneously reducing that element of uncertainty.

types of risks, terrorist organizations want those actions to succeed[9] to maintain the image of their organization, to attract recruits and supporters; and to maintain their level of financial, political, and other support. This desire for success might provide both incentives and disincentives in the pursuit of advanced conventional technologies.

Uncertainties associated with new technologies—their complexities and compatibility with group operations, costs, and benefits—are inevitable, meaning that adopting any new technology has inherent risk associated with it. These risks, which could generate a disincentive for innovation as compared with familiar, tried and true weapons are likely an additional reason for terrorists' historical preference for a small range of attack modes centering on firearms and explosives.[10] Emphasizing their concern about such risks, groups have customized their own technologies with features that allow an operative to verify that the weapon is operational before staging an attack to avoid the chance of self-exposure for no operational payoff. Situations in which training is difficult, e.g., due to security concerns or when limited numbers of weapons are available for testing, will accentuate such problems. When training cannot be carried out, the risk that a weapon cannot be employed as desired can increase considerably.[11]

Conversely, the capabilities designed into these technologies could also make them very attractive from the perspective of an operationally risk-averse terrorist. Although there are risks inherent in new activities, the ease of use of most newer weapons could also bolster the chances

---

[9] It should be noted that what constitutes success in a terrorist operation is not entirely clear and almost certainly differs among groups and contexts. Success could be defined across a broad spectrum based on the objectives of the group in question at a given time. For example, in one operation, success might require achieving a specific outcome; in another, it might simply require demonstration that the group possessed and could use a specific technology or weapon. For additional discussion of the difficulties in measuring success from the point of view of the terrorist, see Jackson, Baker, et al. (2005).

[10] Hoffman (2000).

[11] Although many advanced systems are more complex than currently available analogs, that complexity may not substantially increase the operational risks of adoption. To the extent that a weapon system has been designed to minimize training requirements—such as those described previously that integrate computer assistance and other features—the impact of this last element will be reduced.

that a terrorist operation will be successful. The short time required to set up a GPS-guided mortar system would also significantly reduce any chance detection before the attack.

The experience of PIRA with respect to mortar technologies is illustrative in this regard. Although the group developed considerable skills in designing improvised mortars, PIRA's operational constraints placed limits on its use of the weapons. Because of its desire for its operatives to escape after staging an attack, PIRA took only a "single shot" with its mortars,[12] frequently triggered remotely, and had no opportunity to adjust the aim of its weapons. This meant that PIRA attacks frequently did not strike its intended target or, if they did, did not do so most effectively. Guidance mechanisms being incorporated today into advanced mortar systems would significantly increase the single-shot hit probabilities in such operational situations and, therefore, would strengthen incentives to acquire them. Additionally, the incremental financial cost of the newer mortar systems is not likely to significantly limit their use.

Potential Impacts of Weapon Control Measures

Although control measures would be designed to prevent terrorists from acquiring or using specific advanced weapons of concern, it would be even more preferable if such measures would deter these groups from pursing these weapons at all. The ability of any measure to deter adoption will depend on its capacity to influence the perceived costs, benefits, and risk of pursuing the technology.

The opportunism inherent in many terrorists' weapon choices suggests a cost sensitivity that could provide leverage for shifting groups' decisionmaking on particular weapons. Measures that increase the absolute or perceived costs of these weapons or limit the willingness of other actors, such as sympathetic states—to reduce those costs

---

[12] This is in contrast to a conventional military situation in which error in the aiming of an initial projectile would be observed and corrected for in subsequent volleys. PIRA did produce mortar batteries—so that multiple shells could be fired during an attack—but each tube only fired once, and there was no effort to re-aim the weapons during the operation.

for terrorist groups—could be effective. The sensitivity of any group to measures that increase the costs of advanced weapons will almost certainly depend on its available pool of resources. Efforts to increase the financial costs of these weapons—through interdiction activities intervening in illegal weapon markets, for example—could achieve this goal, though a group with a large resource pool may be willing to simply pay even an inflated cost to gain a desired capability.[13] Such interdiction efforts could have more impact on weapon prices before large numbers of the target weapons are proliferated.

The need to circumvent specific technical controls before use of a weapon would similarly increase prices—both in terms of personnel costs and the technical price of using the weapon. Depending on the design of the countermeasure, that additional cost could influence behavior. Countermeasures that required a distinct effort to break each individual weapon—in contrast to a measure that could be "broken once" and provide access to all similar weapons—would generate a drag on group efforts that could be decisive for resource-constrained organizations.[14]

By their very nature, advanced conventional weapons are "black box" technologies for terrorist organizations. Improvised weapon systems that terrorist organizations manufacture themselves are inherently flexible and transparent. Armed with internal knowledge of their workings and design, a terrorist group can customize them as needed to address operational demands and answer other concerns the group may have about their functioning or effectiveness. This is not the case with the weapon systems discussed in this monograph. Without that level of understanding, the terrorist would need a level of faith in the functionality of the weapon and trust in the sources from which it came. Limitations on groups' abilities to train with weapons—because of scarcity that does not allow them to be used in live-fire training,

---

[13] For example, when al Qaeda represented a centralized and coordinated organization that could concentrate resources from disparate income sources, it is likely it would be less sensitive to such measures than would be a small terrorist cell that had to raise operational funds on its own.

[14] For additional discussion of counterstrategies, see Jackson, Chalk, et al. (2007).

or a lack of "training packages" to enable simulated training—would similarly accentuate this uncertainty by preventing an opportunity for groups to build experience with and trust in the weapon systems.

As discussed previously, a variety of technical strategies can be envisioned to incorporate use controls into these weapons, particularly into those that require access to the GPS for their functioning. Features such as expiration dates, restrictions on weapon use in specific geographic areas, weapons that require periodic updates for continued functionality, or systems that "phone home" with positional information under predefined sets of circumstances—all these factors could influence terrorist groups' cost-benefit risk calculus in ways we would prefer. Controls that cause a weapon to expire could potentially deny terrorists the benefits of their adoption effort, geographic restrictions could result in the failure of an operation when the weapon refuses to operate in the vicinity of the desired target, and systems with the ability to transmit information could compromise the unit's security. Limiting the ability to determine the nature of the countermeasures in a weapon through simple observation would further bolster their effectiveness— e.g., once it was set in the arms depot, the "expiration date" of a weapon could not be read from a mortar unit without specialized equipment or expertise. All such approaches significantly increase the risk inherent in the weapon systems—putting uncertainty inside the "black boxes"— and therefore helping to shape group decisionmaking.

Because nonstate groups frequently rely on the largesse of states (or their rogue officials) to procure advanced weapons, the cost-benefit decisionmaking of terrorist groups is not the only potentially important target of influence. As suggested previously, the resources and capabilities of states may be key to reducing the acquisition costs of these weapons for terrorist organizations. In other words, such resources may be required to circumvent specific use controls that are beyond the means of individual terrorist groups. In contrast to the broad proliferation of MANPADS, where significantly fewer opportunities exist to control their transfer from states to terrorist groups, early intervention, before the broad distribution of new systems, could preserve more options. Traceability concepts such as ways to detect by whom or where a unit's use controls were hacked, for example, could help provide a "return address" to the state sponsors that supplied these weapons to terrorist groups and enabled their use.

The effectiveness of specific use control strategies will also depend on where terrorist groups are obtaining weapons. Because of the added resource costs they impose on the use of a weapon and the risks they may pose for operational effectiveness, the potential deterrent effect of use controls is strongest when a group must confront and defeat the controls on its own. If a group gains a weapon through theft, for example, it will have only its own technical capabilities and resources to bring to bear on breaking the control and may, therefore, be deterred by the costs of doing so. If external sources are providing weapons to the group, the situation could change considerably. Sophisticated providers—e.g., sympathetic states—are likely to have the technical capabilities to break use controls if they chose to do so. This would remove any added cost or risk to the terrorist group and would neutralize any deterrent to terrorist use of the weapons, provided that the terrorist organization has sufficient confidence and trust that its state sponsor would indeed effectively and completely defeat the weapon's control mechanisms. Other providers, such as sellers in the arms market, might or might not have the capabilities required to defeat all relevant use controls. It is also likely that the confidence a terrorist group would have in the skills of arms market dealers (and their motives for doing so) might be considerably lower than its confidence in a sympathetic state. As a result, this could preserve at least a partial deterrent effect for some or all groups, even if weapon suppliers can circumvent control mechanisms.

Conclusions

Though the capabilities provided by many advanced conventional weapons will almost certainly make them attractive to terrorists, their acquisition and use by extremist groups will depend on how these weapons' relative costs, benefits, and risks are perceived. Those perceptions therefore provide a point at which technical and other use controls could influence decisionmaking and deter acquisition and use of the weapons.

The diversity of terrorist behavior and organizations makes it difficult to predict with any certainty the specific effects of any individual use control approach. Differences among groups will result in some being willing to pay acquisition costs that others will not. Differences in technical capabilities will mean that some groups may possess the capabilities to circumvent controls that will prove insurmountable to others. The heterogeneity inherent in the problem means that it is unlikely that precise quantitative measures could be developed to assess the efficacy of use control strategies.

Since cross-organizational learning does occur among terrorist groups, it is particularly important to avoid the first use of a new system. Once one terrorist group has demonstrated the successful use of some new weapon system, it becomes much more likely that other groups will end up sharing this capability.[15]

In the absence of certainty, consideration of controls must rest on assessment of how the potential costs of terrorist acquisition and use compared to the costs involved in responding to the threat. As we have seen, the use of some of these technologies could provide a powerful force multiplier for terrorist groups. Once acquisition of such a weapon by a terrorist organization has been confirmed, protection and hardening strategies for a wide range of targets would be rendered less effective or obsolete, requiring a significant response. Steps that might be taken after their acquisition and use by security organizations or potential targeted sites—reinforcement of security forces around key installations, fortification of sensitive targets—would be costly and are primarily parity strategies that would still leave major advantages and capabilities in the hands of the adversary. Against such costs, investments that seek to deter terrorist organizations from acquiring and using these weapons or that impose limits on the way they could be used appear to be an increasingly attractive approach within a broad-based effort to counter terrorist activity and limit the threat posed by extremist groups.

---

[15] Jackson, Chalk, et al. (2007).

CHAPTER SIX

Observations and Implications

This research started with two basic questions, which can now be answered:

- What difference would it make if terrorists could use advanced conventional weapons in their attacks?
- What could the United States do to reduce this threat?

Most advanced conventional weapons are not discussed in depth in this monograph because they do not appear to be particularly attractive to terrorists. In many cases, terrorists already have roughly equivalent weapons at their disposal, and incremental improvements will not significantly increase their attack capabilities. For example, although an improved explosive might enable terrorists to make a truck bomb smaller, existing truck bombs have been sufficient for most terrorists' needs and desired targets. Therefore, acquisition of some types of advanced weapons by terrorists would not significantly change the balance of capabilities between terrorists and security forces.

But if terrorists gained use of some advanced conventional weapons, the competition would change significantly. This book has identified five types of advanced conventional weapons that would, in some sense, "change the game" between terrorists and security forces. These were precision indirect fire systems, improved squad-level weapons of several types, sniper rifles and instrumentation, long-range antitank missiles, and large limpet mines.

Precision indirect fire systems—various designs for improved mortars—could enable terrorists to attack quickly and precisely from a distance, with no line of sight or observer near the target. Many new classes of targets would be at great risk, from public figures at some outdoor event to crowds of people at an outdoor sporting event. Additionally, the ability of these advanced systems to strike from a distance could easily allow the terrorists to escape and strike again.

Improved squad-level weapons could provide terrorists assaulting a guarded location with the ability to overwhelm an unprepared security force. In principle, this could place many guarded locations at risk, such as nuclear power plants or chemical plants.

Sniper rifles and their associated support equipment could allow assassination attempts from unexpectedly long ranges. Although such rifles have many other, countermaterial capabilities, this appears to be the most worrisome terrorist application. This would be a threat, for example, for those individuals afforded personal protection, as by the Secret Service.

Long-range, antitank weapons can destroy essentially *any* vehicle from a few kilometers away. Some can also be used to attack fixed locations, such as a speaking stand or a small building. Although existing versions of these systems already could execute these attacks, newer systems are increasingly easy to use, allowing terrorists who may have limited opportunity to train or to test these weapons to still carry out such an attack.

Large limpet mines, together with the associated underwater platforms for transporting the mines, could threaten even larger ships with catastrophic damage and sinking. For terrorist groups, such mines would circumvent the inspection of cargo for explosives. Cruise ships and oceangoing ferries that carry large numbers of passengers would appear to be the most attractive targets for such mines, though almost all ships would be vulnerable.

That group of five are the most worrisome advanced conventional weapons that were identified in this book. Fortunately, there are some actions the United States can take to reduce all of these threats.

The most important response is simple awareness. If security forces are aware of the new capabilities that terrorists might possess, they can adjust their tactics or equipment to compensate. In the case of one weapon system above—advanced sniper rifles—this response will be the only one available. Sniper rifles and their support equipment are already widely available within the United States, so control seems implausible. Consequently, security forces concerned with sniper attacks must extend their control over all lines of sight out to about 2 km. That measure would largely obviate the effect of these weapons for those few officials afforded such protection, such as a president or presidential candidates.

These same security forces must also become aware of several other weapons mentioned above. Antitank weapons are lethal, line-of-sight weapons reaching well over 2 km. The security forces also need to realize that non–line-of-sight weapons, such as the mortars described earlier, will soon allow very long-range, precise attacks on targets at known locations. This awareness should allow the protective services to make the opportunities for terrorists to use both of these weapons rarer and, above all, more unpredictable.

Guard forces at many places need to be aware of the capabilities that new, squad-level weapons would provide to an assault force attacking them. The addition of precise, indirect-fire grenades should drive a greater concern with overhead cover. New RPGs, thermobaric warheads, and short-range antitank weapons will change the fortification needed at any strong points. Even the small-arms ammunition now available makes a difference, changing the protection that should be expected from simple personal armor.

Cruise ship and ferry operators must become aware of the potential of large limpet mines. This should motivate hull inspections before leaving port, either using divers or robotic undersea vehicles. Although such inspections might trigger some mines, at the worst, the ship would sink at dockside—a decidedly less dramatic result with less loss of life.

Awareness of the threat posed by these advanced conventional weapons does not mitigate their threat entirely. Fortunately, with the exception of the sniper rifles and their support equipment, all these systems are legally limited to established militaries. This means the existing sets of export control regimes and domestic laws limiting munitions will apply. This limitation will certainly slow the diffusion of these systems to terrorist groups. But, as with previous weapons, some erosion of these procedural controls over time is likely to be inevitable.

The danger created by that erosion could motivate either or both of two international efforts. The first effort would be to place selected advanced weapon systems under tighter procedural controls. The second effort would be to impose technical controls on these weapons where feasible. Both efforts would involve international agreements among the producing nations, which takes time and effort both within the United States to agree to start the diplomatic discussions and then also internationally, to forge an agreement among nations.

The most obvious type of advanced conventional weapons that could motivate these efforts would be the precise, indirect-fire systems—the advanced mortars. These systems enable a wide variety of new, unfamiliar terrorist attacks. The attacks would not require complex training or preparation, unlike an assault of a guarded facility. Some of these attacks, such as those against outdoor crowds or outdoor speaking events, would be difficult for even fully aware security forces to counter. Finally, several terrorist groups have experience with existing mortars and have exploited them for attacks, unlike other types of advanced conventional weapons such as long-range, antitank weapons that have seldom been used. Only the advanced mortar systems combine all these factors.

Additionally, among these weapon systems, only the advanced mortar systems seem plausible candidates for technical control measures. First, the most worrisome systems are all still in development or even earlier, in a design phase. This means that integrating control measures may still be possible without great effort. Even more important though, these most worrisome systems rely upon a satellite navigation system, typically the GPS, and some sort of IMU to fly to their target. This reliance on GPS means that the systems could be designed to depend upon a secure component that integrates a GPS receiver and an IMU. In principle, such a secure component would then allow various technical controls to be implemented—the need for a cryptographic key to unlock the system when it is checked out for use; a time limit on the period of use; and geographic limits on the area where the system would function. All of these controls would make the weapon system much less dependable for a terrorist group and so significantly reduce its probable interest in the weapons.

Achieving an international agreement to impose additional limits on these advanced mortar systems is very likely to be a time-consuming process. The similar discussions on limiting MANPADS was a subject of discussion under the Wassenaar Arrangement in 1998, but agreement on both procedural and very general technical controls was not reached until 2003.[1] To implement controls before these systems become dispersed means that the United States must act soon, or the window of opportunity will close.

If the United States chooses to pursue additional controls on advanced mortars, the first two steps are clear. One is to begin diplomatic discussions with the key producer nations so they begin thinking about the potential of terrorist groups to use these systems. The other is to commission a detailed technical study of the architecture needed to implement any technical controls. The existing technical modules and architecture may be sufficient, but they might need to be modestly expanded to include the control functions described here.

The Department of Homeland Security can play a key role in both of these steps. For the first step, the department has the primary responsibility for considering terrorist attacks. It could use that role to push within the interagency process for starting diplomatic discussions. This may also entail changes in the interagency system to permanently include the Department of Homeland Security on the interagency panels considering arms exports. For the second step, the department could itself directly fund such a study, probably involving the National Security Agency.

Although it appears that there is sufficient time to negotiate and develop meaningful controls, that opportunity can be lost if the United States does not begin the process now. Missing this opportunity would reduce the controls on these mortars to the existing procedural ones for military systems, which, in turn, would increase the burden on security forces to plan around these attacks. Although that may be a sufficient

---

[1] Wassenaar Arrangement on Export Controls for Conventional Arms and Dual-Use Goods and Technologies (1998, 2003).

response for the other weapon systems we have analyzed, it appears to us to be insufficient for limiting the threat from the GPS-guided advanced mortars. For this reason, we trust that this monograph will spur the United States on to take the first steps toward such controls.

Bibliography

5.11 Tactical Series, "Accessories," undated Web page. As of February 14, 2007: http://www.511tactical.com/index.asp?dlrID=511&dept=7

Adorama, "Leica Rangemaster 1200 Scan Mode Laser Rangefinder with 1,200 Yd. Range, Black—USA," undated Web page. As of February 13, 2007: http://www.adorama.com/LCLRF12SM.html

Alliant Techsystems, "25mm Airburst Weapon System," undated Web page.

American Technologies Network Corporation, "ATN Aries MK8700 Night Vision Weapon Sight," undated. As of July 19, 2006: http://www.atncorp.com/NightVision/NightVisionWeaponSights/ATNAriesMK8700

"Ammunition for 81mm M29 and M29A1 Mortars," *Jane's Infantry Weapons*, October 15, 2003.

Anderson, Ross, *Security Engineering: A Guide to Building Dependable Distributed Systems*, New York: Wiley, 2001.

Anderson, Ross, Mike Bond, Jolyon Clulow, and Sergei Skorobogatov, "Cryptographic Processors—A Survey," *IEEE Proceedings*, April 2005, pp. 100–116.

"Anti-Armor Missiles," *Army Technology*, August 2005.

"Anti-Helicopter Mine Displayed at HEMUS-96," *Jane's Intelligence Review*, EUROPE/CIS, Vol. 3, No. 8 August 1, 1996.

"Anti-Helicopter Mines," *Jane's Mines and Mine Clearance*, July 15, 2005.

ATK—*see* Alliant Techsystems.

"ATK XM25 Grenade Launcher for Future Industry: Will It Fly?" *Defense Review*, May 2006.

Aurora Tactical, "Why Night Vision?" undated Web page.

BAE Systems, "BAE Systems to Develop 60mm Mortar Guidance Systems," February 14, 2005. As of August 21, 2007:
http://www.baesystems.com/Newsroom/NewsReleases/2005/press_14022005.html

"BAE Systems Designs Precision Seeker for Mortar Rounds," *SpaceDaily*, January 31, 2005. As of July 31, 2006:
http://www.spacedaily.com/news/gps-05q.html

Baker, John C., Beth E. Lachman, David R. Frelinger, Kevin M. O'Connell, Alexander C. Hou, Michael S. Tseng, David T. Orletsky, and Charles W. Yost, *Mapping the Risks: Assessing the Homeland Security Implications of Publicly Available Geospatial Information*, Santa Monica, Calif.: RAND Corporation, MG-142-NGA, 2004. As of February 14, 2007:
http://www.rand.org/pubs/monographs/MG142/index.html

"Balad Airbase," *GlobalSecurity.org*, undated. As of July 18, 2006:
http://www.globalsecurity.org/military/world/iraq/balad-ab.htm

Baptista, Rui, "The Diffusion of Process Innovations: A Selective Review," *International Journal of the Economics of Business*, Vol. 6, No. 1, 1999, pp. 107–130.

Bayles, Fred, "Threat Is 'No Longer Theoretical,'" *USA Today*, August 13, 2003. As of February 13, 2007:
http://www.usatoday.com/news/nation/2003-08-13-insidemissile-usat_x.htm

BBC News, "1981: President Reagan Is Shot," *On This Day*, undated Web page. As of February 13, 2007:
http://news.bbc.co.uk/onthisday/hi/dates/stories/march/30/newsid_2530000/2530913.stm

Bell, J. Bowyer, *The Gun in Politics: An Analysis of Irish Political Conflict, 1916–1986*, New Brunswick, N.J.: Transaction Books, 1987.

———, *The Secret Army: The IRA*, Dublin, Ireland: Poolbeg, 1998.

Biass, Eric H. "Drilling the Darkness," *Armada International*, June 2004a.

———, "What Is So Special?" *Armada International*, June 2004b.

Bischer, Greg, "Precision Guided Mortar Munition (PGMM) XM395," briefing at 1999 International Infantry and Small Arms Symposium, June 22, 1999. As of July 27, 2006:
http://www.dtic.mil/ndia/infantry/bischer.pdf

Blair, Bruce, *The Logic of Accidental Nuclear War*, Washington, D.C.: The Brookings Institution Press, 1993.

"BLU-118/B Thermobaric Weapon," *GlobalSecurity.org*, November 2005. As of July 31, 2006:
http://www.globalsecurity.org/military/systems/munitions/blu-118.htm

Bond, Mike, and Ross Anderson, *API-Level Attacks on Embedded Systems,* University of Cambridge, May 2, 2001. As of February 13, 2007: http://www.cl.cam.ac.uk/~rja14/Papers/API-Attacks.pdf

Bracken, Paul, *The Command and Control of Nuclear Forces,* Yale University Press, 1983.

Caffera, Paul J., "The Vexing Problem of Protecting Airliners from MANPADS," *Aircraft Survivability,* Spring 2003.

CheyTac, undated homepage. As of February 13, 2007: http://www.cheytac.com/

———, *Advanced Ballistic Computer,* version 1.96, *User's Guide,* February 4, 2004. As of February 13, 2007: http://www.cheytac.com/ABC%20v1.96%20UserGuide%20FINAL.pdf

CheyTac, "CheyTac® Intervention™: CheyTac Long Range Rifle System, Intervention™ Tactical System," September 2001–November 2006. As of February 13, 2007: http://www.cheytac.com/White%20Paper2007.pdf

Cilli, Matthew V., and Gregory Bischer, "Precision Guided Mortar Munition (PGMM)," briefing, Fort Belvoir, Va.: 35 Annual Gun and Ammunition Symposium, Defense Technical Information Center, May 1–4, 2002.

CNN, "Rabin's Alleged Killer Appears in Court," *CNN World News,* November 7, 1995. As of February 13, 2007: http://www.cnn.com/WORLD/9511/rabin/amir/11-06/index.html

Combat Games, "Activities: Sniper School," undated Web page. As of February 13, 2007: http://www.ukpaintballgames.com/activities-sniper.html

"Coriolis Effect," *The Columbia Electronic Encyclopedia,* 6th ed., 2007. As of July 31, 2006: http://www.infoplease.com/ce6/weather/A0813558.html

Cotter, Donald, "Peacetime Operations: Safety and Security," in Ashton B. Carter, John D. Steinbruner, and Charles A. Zraket, eds., *Managing Nuclear Operations,* Washington, D.C.: Brookings Institution Press, 1987.

Daniel, Eric, "Sniping Made Easy: The TRAP System," *Military.com,* 2004. As of February 13, 2007: http://www.military.com/soldiertech/0,14632,Soldiertech_TRAP,,00.html

Davidc, "Metal Storm Demos Advanced Individual Combat Weapon (AICW) Prototype (Photo)," *Defense Review,* August 31, 2005. As of February 13, 2007: http://www.defensereview.com/modules.php?name=News&file=article&sid=781

Defense Technical Information Center, undated homepage. As of February 13, 2007:
http://www.dtic.mil/

Donovan, John, "XM29/XM25 Spiral Development Strategy," September 17, 2003, *The Donovan.com*. As of February 13, 2007:
http://www.thedonovan.com/John/XM29Briefedited.ppt

Forest, James J. F., ed., *The Making of a Terrorist: Recruitment, Training, and Root Causes*, Westport, Conn.: Praeger Security International, 2005.

Gander, Terry J., "Lighter, Smarter, and on the Move," *Armada International*, November 2000. As of July 31, 2006:
http://www.armada.ch/00-6/001.htm

General Dynamics, "XM307: 25mm Airbursting Weapon System," 2005. As of July 20, 2006:
http://www.gdatp.com/Products/PDFs/XM307.pdf

Generation Airsoft, "KSC Glock 18c," undated Web page. As of February 13, 2007:
http://www.generationairsoft.com/wpn/GAS_KSC_g18c.htm

Geraghty, Tony, *The Irish War: The Hidden Conflict Between the IRA and British Intelligence*, Baltimore, Md.: Johns Hopkins University Press, 2000.

Giammario, "The *HMS Valiant* British Battleship in Alexandria Harbour 1941," *Abysso*, undated. As of July 21, 2006:
http://www.sportesport.it/wrecksEG014.htm

GizMag.com, "XM25 Prototypes in Testing—500% Lethality Increase Over Existing Weapon Systems," May 27, 2005. As of February 13, 2007:
http://www.gizmag.com/go/4081/

Gourley, Scott, "XM32 Lightweight Handheld Mortar Ballistic Computer," *Army*, September 2004.

"Gran 120mm Guided Mortar Bomb: U.S. Developments," *Defense Update*, December 2004. As of July 31, 2006:
http://www.defense-update.com/products/g/Gran.htm

Hansen, Chuck, *U.S. Nuclear Weapons: The Secret History*, Arlington, Tex.: Aerofax, 1988.

Haugen, Michael, "The Truth About Mil Dots," *Boomershoot*, 2001. As of August 2, 2006:
http://www.boomershoot.org/general/TruthMilDots.htm

Hoffman, Bruce, "Change and Continuity in Terrorism," after-dinner address delivered at the Terrorism and Beyond: The 21st Century Conference, cosponsored by the Oklahoma City National Memorial Institute for the Prevention of Terrorism and the RAND Corporation, April 17, 2000. As of February 13, 2007: http://www.terrorisminfo.mipt.org/hoffman-ctb.asp

Horus Vision, "Software," undated Web page. As of February 13, 2007: http://www.horusvision.com/software.html

————, *ATrag User's Manual: Palm and PocketPC*, version 3.6x for 1P, 2P, and MP, San Bruno, Calif.: Horus Vision, January 1, 2007. As of February 15, 2007: http://www.horusvision.com/down/Horus070101-ATrag36x.pdf

Huang, Andrew, "Keeping Secrets in Hardware: The Microsoft XBox™ Case Study," AI Memo 2002-008, Massachusetts Institute of Technology Artificial Intelligence Laboratory, May 26, 2002.

————, *Hacking the Xbox: An Introduction to Reverse Engineering*, San Francisco, Calif.: No Starch Press, 2003.

Hunter, Bob, "France Says It with Bombs," *Eye Weekly*, July 13, 1995. As of February 13, 2007: http://www.eye.net/eye/issue/issue_07.13.95/NEWS/env0713.php

IAI—*see* Israel Aircraft Industries.

"Infantry: 81mm L16 Mortar," *The British Army*, undated Web page. As of July 31, 2006: http://www.armedforces.co.uk/army/listings/l0098.html

Israel Aircraft Industries, "Fireball—Smart Mortar Bomb Weapon System," undated Web page. As of August 1, 2006: http://www.iai.co.il/Default.aspx?docID=30914&FolderID=31015&lang=en

————, "Fireball—Precision Mortar Munition," *Defense Update*, February 2004. As of July 22, 2006: http://www.defense-update.com/products/f/Fireball.htm

Jackson, Brian A., "Provisional Irish Republican Army," in Brian A. Jackson, John C. Baker, Peter Chalk, Kim Cragin, John V. Parachini, and Horacio R. Trujillo, *Aptitude for Destruction*, Vol. 2: *Case Studies of Organizational Learning in Five Terrorist Groups*, Santa Monica, Calif.: RAND Corporation, MG-332-NIJ, 2005. As of February 14, 2007: http://www.rand.org/pubs/monographs/MG332/

Jackson, Brian A., John C. Baker, Peter Chalk, Kim Cragin, John V. Parachini, and Horacio R. Trujillo, *Aptitude for Destruction*, Vol. 1: *Organizational Learning in Terrorist Groups and Its Implications for Combating Terrorism*, Santa Monica, Calif.: RAND Corporation, MG-331-NIJ, 2005. As of February 14, 2007: http://www.rand.org/pubs/monographs/MG331/index.html

Jackson, Brian A., Peter Chalk, Kim Cragin, Bruce Newsome, John V. Parachini, William Rosenau, Erin M. Simpson, Melanie Sisson, and Donald Temple, *Breaching the Fortress Wall: Understanding Terrorist Efforts to Overcome Antiterrorism Technologies*, Santa Monica, Calif.: RAND Corporation, MG-481-DHS, 2007. As of February 14, 2007:
http://www.rand.org/pubs/monographs/MG481/

Jane's Underwater Warfare Systems, June 1, 2005.

Kennedy, Harold, "Back to the Drawing Board: Army Rewrites Small Arms Plans," *National Defense*, July 2005. As of July 22, 2006:
http://www.nationaldefensemagazine.org/issues/2005/Jul/Back_to.htm

Laser-King Companies, "Thermal Imaging Devices," undated. As of November 8, 2006:
http://www.laser-king.com/thermal_imaging.html

"M395 Precision-Guided Mortar Munition (PGMM)," *GlobalSecurity.org*, April 2005. As of July 23, 2006:
http://www.globalsecurity.org/military/systems/munitions/pgmm.htm

MBDA, undated homepage. As of February 13, 2007:
http://www.mbda-systems.com

MBDA Missile Systems, "Press Kit," undated Web page.

———, "ALFO: Lightweight Fibre Optic Guided Missile," May 2003.

"MBT LAW Light Anti-Tank Weapon, United Kingdom," *army-technology.com*, undated. As of July 28, 2006:
http://www.army-technology.com/projects/mbt_law/

Metal Storm, undated homepage. As of February 13, 2007:
http://www.metalstorm.com

———, "2002 NDIA SO/LIC Symposium and Exhibition: The War on Terrorism," February 8, 2002. As of February 13, 2007:
http://www.metalstorm.com/clientuploads/news/1presentations/SO_LIC_Brief_8_Feb_2002.pdf

———, "Electronic Prototype," 2004a. As of August 4, 2006:
http://www.metalstorm.com/04_electronic_prototype.html

———, "Internal Build Program," May 21, 2004b. As of February 13, 2007:
http://www.metalstorm.com/clientuploads/news/1presentations/MS%20Build%20Review%20AGM%202004%20Website.pdf

———, "VLe," video, 2004c.

"Mini Submarines and Special Forces Pose Maximum Threat," *Jane's International Defense Review*, June 1998.

Modern Ballistics, undated homepage. As of February 13, 2007:
http://www.modernballistics.com

"Mortar Systems," *Picatinny, Home of American Fire Power*, undated Web page. As
of July 28, 2006:
http://www.pica.army.mil/PicatinnyPublic/products_services/products03.asp

"Mortar (Weapon)," *Wikipedia*, undated Web page. As of July 28, 2006:
http://en.wikipedia.org/wiki/Mortar_%28weapon%29

"Mortars," *FAS Military Analysis Network*, September 12, 1998. As of February 13,
2007:
http://www.fas.org/man/dod-101/sys/land/mortars.htm

Murdoc, "Smart Grenade Launcher," *Murdoc Online*, February 20, 2004. As of
July 20, 2006:
http://www.murdoconline.net/archives/001036.html

Nappert, L., Y. Champagne, and Y. Taillon, "Feasibility of an Eye-Safe Laser-
Based Crosswind Velocity Measurement System for Sniper Rifles," briefing,
Atlantic City, N.J.: International Infantry and Joint Services Small Arms Systems
Section Symposium, Exhibition and Firing Demonstration, May 13–16, 2002. As
of February 13, 2007:
http://www.dtic.mil/ndia/2002infantry/nappert.pdf

National Defense Magazine, August 2004. As of July 18, 2006:
http://www.nationaldefensemagazine.org/issues/2004/Aug/Army_Will.htm

National Memorial Institute for the Prevention of Terrorism, and RAND
Corporation, *RAND-MIPT Terrorism Incident Database*, Oklahoma City, Okla.:
Oklahoma City National Memorial Institute for the Prevention of Terrorism,
2002–.

National Research Council, *Safety and Security of Commercial Spent Nuclear Fuel
Storage: Public Report*, Washington, D.C.: National Academies Press, 2006.

Nielsen Kellerman, "Kestrel® Wind and Weather Instruments," undated Web
page. As of February 13, 2007:
http://www.nkhome.com/ww/wwindex.html

"Nikon Rangefinders," Southern C&E Professional Duty Products, undated Web
page.

O'Callaghan, Sean, *The Informer*, London, UK: Corgi Books, 1999.

Pasadena Tournament of Roses, "Rose Bowl Game FAQs," undated Web page. As
of February 13, 2007:
http://www.tournamentofroses.com/rosebowlgame/gamefaqs.asp

Pascua, Daniel, "XM984 120mm Mortar Cartridge Extended Range DPICM," Picatinny Arsenal, N.J.: U.S. Army Tank-Automotive and Armaments Command, May 14, 2002. As of July 28, 2006:
http://www.dtic.mil/ndia/2002infantry/pascua.pdf

"Precision Guided Mortar Munition (PGMM)," undated briefing, author unknown. As of February 13, 2007:
http://www.dtic.mil/ndia/ammo/cilli.pdf

Precision Remotes, undated homepage. As of August 2, 2006:
http://precisionremotes.com

"RAND-MIPT Terrorism Knowledge Base," established April 1995. As of July 22, 2006:
http://www.tkb.org/

Raytheon Company, "Successful Flight Test of GPS-Guided 20 Artillery Projectile Puts Raytheon-Bofors Excalibur Closer to Fielding," press release, September 26, 2005.

"Red Dot Sight," *Wikipedia*, July 2006. As of August 3, 2006:
http://en.wikipedia.org/wiki/Red_dot_sight

Rice, Barry, "Remote Sensing and Invasive Species: A History, Glossary, and More!" *The Nature Conservancy Global Invasive Species Initiative*, November 2004. As of February 13, 2007:
http://tncweeds.ucdavis.edu/remotesensing.html

Ripley, Tim, and E. H. Biass, "Mortars for the 21st Century, *Armada International*, May 1998. As of February 13, 2007:
http://www.armada.ch/98-5/001.htm

Rogers, Everett M., *Diffusion of Innovations*, New York: The Free Press, 1995.

Saab, "STRIX," *Saab Products*, undated Web page.

Schneier, Bruce, *Applied Cryptography: Protocols, Algorithms, and Source Code in C*, 2nd ed., New York: Wiley, 1996.

Shipunov, Arkady, Victor Babichev, Alexander Ignatov, and Vladimir Rabinovich, "Gran Mobile High-Precision Weapon System," *Military Parade,* March–April 2003, pp. 54–56. As of July 31, 2006:
http://www.milparade.com/2003/56/54-56.pdf

"Shoot-Out at Blackwater," *Armed Forces Journal*, 2005.

Sniper Country, undated homepage. As of February 13, 2007:
http://www.snipercountry.com/

"STRIX Precision Guided 120mm Mortar Launched Weapon," *Defense Update*, January 2004. As of July 31, 2006:
http://www.defense-update.com/products/s/strix.htm

Sundra, Jon, "High-Tech Optics Feed Customers' Desire for Gizmos—Riflescopes and Binoculars," *Shooting Industry*, June 1999.

Talley Defense Systems, undated homepage. As of February 13, 2007:
http://www.talleyds.com/

Teaching Terror, undated homepage. As of February 13, 2007:
http://www.teachingterror.com

Tiron, Roxana, "Army Will Boost Supply of Small Cal Ammo, Weapons," *National Defense*, August 2004. As of February 13, 2007:
http://www.nationaldefensemagazine.org/issues/2004/Aug/Army_Will.htm

Trijicon, "Bindon Aiming Concept: The Nature of Binocular Sighting," undated Web page. As of February 13, 2007:
http://www.trijicon-inc.com/aiming.cfm

"Trijicon Accupoint Scopes," Wholesale Hunter, undated Web page.

Trueman, Chris, "The Cockleshell Heroes of 1942," *History Learning Site*, undated Web page. As of July 21, 2006:
http://www.historylearningsite.co.uk/cockleshell_heroes_of_1942.htm

Trusted Computing Group, *TCG Specification Architecture Overview Specification*, 2004.

"Type 35, 120mm Mortar," *Sino Defence*, undated Web page.

"Type 64, 120mm Mortar," *Sino Defence*, undated Web page.

"Type 86 120mm Mortar," *Sino Defence*, undated Web page.

Urban, Mark, *Big Boys' Rules: The Secret Struggle Against the IRA*, London, UK: Faber and Faber, 1992.

U.S. Army, *America's Army*, undated computer game.

———, "They Have the Will . . . We Have the Way," *PEO Soldier Portfolio*, October 2005. As of February 15, 2007:
http://peosoldier.army.mil/docs/peoportfolio06.pdf

U.S. Department of the Army, "Sniper Training and Employment," *U.S. Army Field Manual*, Washington, D.C.: U.S. Department of the Army, FM 23-10, August 17, 1994. As of February 13, 2007:
https://atiam.train.army.mil/soldierPortal/atia/adlsc/view/public/9504-1/fm/23-10/toc.htm

U.S. Department of Defense, "Exhibit R-2a, RDT&E Project Justification," February 2003. As of February 13, 2007:
http://www.dod.mil/comptroller/defbudget/fy2004/budget_justification/pdfs/rdtande/OSD_RDTE/Budget_Activity_3/07_0603122__R2A__Feb_2003.pdf

U.S. Department of Homeland Security, "Fact Sheet: U.S. Department of Homeland Security Programs Countering Missile Threats to Commercial Aircraft," August 25, 2004. As of July 18, 2006:
http://www.dhs.gov/xnews/releases/press_release_0497.shtm

U.S. Department of State, "Advanced Conventional Weapons," undated Web page.

U.S. Developments, "Precision Munitions for 120 mm Mortars," *Defense Update*, December 24, 2004. As of July 31, 2006:
http://www.defense-update.com/features/du-1-04/advanced-mortar-munition.htm

U.S. Joint Chiefs of Staff, *Department of Defense Dictionary of Military and Associated Terms*, Washington, D.C.: Joint Chiefs of Staff, Joint Publication 1-02, April 12, 2001, as amended through January 5, 2007. As of February 13, 2007:
http://purl.access.gpo.gov/GPO/LPS14106

Wassenaar Arrangement on Export Controls for Conventional Arms and Dual-Use Goods and Technologies, "Public Statement—3 December 1998," December 3, 1998. As of February 13, 2007:
http://www.wassenaar.org/publicdocuments/public031298.html

———, "Elements for Export Controls of Man-Portable Air Defence Systems (MANPADS)," adopted at the 2003 plenary. As of July 24, 2006:
http://www.wassenaar.org/2003Plenary/MANPADS_2003.htm

"XM25 Individual Airburst Weapon System," *Program Executive Office Soldier*, October 2006. As of February 13, 2007:
https://peosoldier.army.mil/factsheets/SW_IW_XM25.pdf

"XM25 Prototype in Testing," *Defense Review*, 2005. As of July 26, 2006:
http://www.defensereview.com/modules.php?name=News&file=article&sid=730

Yoo, John, "Tactical Fuze Design for the XM984 DPICM Mortar Cartridge," San Antonia, Tex.: 46th Annual NDIA Fuze Conference, April 29–May 1, 2002. As of July 28, 2006:
http://www.dtic.mil/ndia/2002fuze/yoo.pdf